AF677072

Acc No. 28812 9760

231 Current Topics in Microbiology and Immunology

Editors

R.W. Compans, Atlanta/Georgia
M. Cooper, Birmingham/Alabama
J.M. Hogle, Boston/Massachusetts · Y. Ito, Kyoto
H. Koprowski, Philadelphia/Pennsylvania · F. Melchers, Basel
M. Oldstone, La Jolla/California · S. Olsnes, Oslo
M. Potter, Bethesda/Maryland · H. Saedler, Cologne
P.K. Vogt, La Jolla/California · H. Wagner, Munich

Springer
Berlin
Heidelberg
New York
Barcelona
Budapest
Hong Kong
London
Milan
Paris
Santa Clara
Singapore
Tokyo

Leukocyte Integrins in the Immune System and Malignant Disease

Edited by B. Holzmann and H. Wagner

With 40 Figures

Springer

Professor Bernhard HOLZMANN, M.D.
Professor Hermann WAGNER, M.D., Ph.D.

Technical University of Munich
Institute of Medical Microbiology,
Immunology, and Hygiene
Trogerstr. 9
D-81675 Munich
Germany

Cover illustration: Cell adhesion receptors of the integrin family are important elements of cellular communication. The capacity of integrins to bind ligand is rapidly regulated by cellular signaling events that alter integrin conformation and clustering. Upon recognition and binding of cognate ligands integrins may also function as signal transduction devices that affect cellular responses to environmental signals.

Cover design: design & production GmbH, Heidelberg

ISSN 0070-217X
ISBN 3-540-63609-9 Springer-Verlag Berlin Heidelberg New York

This work is subject to copyright. All rights are reserved, whether the whole or part of the material is concerned, specifically the rights of translation, reprinting, reuse of illustrations, recitation, broadcasting, reproduction on microfilm or in any other way, and storage in data banks. Duplication of this publication or parts thereof is permitted only under the provisions of the German Copyright Law of September 9, 1965, in its current version, and permission for use must always be obtained from Springer-Verlag. Violations are liable for prosecution under the German Copyright Law.

© Springer-Verlag Berlin Heidelberg 1998
Library of Congress Catalog Card Number 15-12910
Printed in Germany

The use of general descriptive names, registered names, trademarks, etc. in this publication does not imply, even in the absence of a specific statement, that such names are exempt from the relevant protective laws and regulations and therefore free for general use.

Product liability: The publishers cannot guarantee the accuracy of any information about dosage and application contained in this book. In every individual case the user must check such information by consulting other relevant literature.

Typesetting: Scientific Publishing Services (P) Ltd, Madras

SPIN: 10630629 27/3020 – 5 4 3 2 1 0 – Printed on acid-free paper

List of Contents

C. Morimoto, S. Iwata, and K. Tachibana
VLA-4-Mediated Signaling 1

N. Wagner and W. Müller
Functions of α4-β7-Integrins in Hematopoiesis, Lymphocyte Trafficking and Organ Development 23

W. Kolanus and L. Zeitlmann
Regulation of Integrin Function by Inside-Out Signaling Mechanisms. 33

B. Engelhardt
The Role of α4-Integrins in T Lymphocyte Migration Into the Inflamed and Noninflamed Central Nervous System 51

S.A. Michie, H.-K. Sytwu. J.O. McDevitt, and X.-D. Yang
The Roles of α4-Integrins in the Development of Insulin-Dependent Diabetes Mellitus 65

M. Isobe, J. Hori, and J. Suzuki
Immunosuppression by Blocking α4-Integrins/VCAM-1 Adhesion 85

A.R. Thomsen, A. Nansen, and J.P. Christensen
Virus-Induced T Cell Activation and the Inflammatory Response 99

B. Holzmann, U. Gosslar, and M. Bittner
α4 Integrins and Tumor Metastasis 125

D. Naor, R. Vogt Sionov, M. Zahalka, M. Rochman, B. Holzmann, and D. Ish-Shalom
Organ-Specific Requirements for Cell Adhesion Molecules During Lymphoma Cell Dissemination 143

M.M. Zutter and S.A. Santoro
The Ups and Downs of $\alpha_2\beta_1$-Integrin Expression: Contributions to Epithelial Cell Differentiation and the Malignant Phenotype 167

Subject Index 187

List of Contributors

(Their addresses can be found at the beginning of their respective chapters.)

Bittner, M. 125

Christensen, J.P. 99

Engelhardt, B. 51

Gosslar, U. 125

Holzmann, B. 125, 143

Hori, J. 85

Ish-Shalom, D. 143

Isobe, M. 85

Iwata, S. 1

Kolanus, W. 33

McDevitt, J.O. 65

Michie, S.A. 65

Morimoto, C. 1

Müller, W. 23

Nansen, A. 99

Naor, D. 143

Rochman, M. 143

Santoro, S.A. 167

Suzuki, J. 85

Sytwu, H.-K. 65

Tachibana, K. 1

Thomsen, A.R. 99

Vogt sionov, R. 143

Wagner, N. 23

Yang, X.-D. 65

Zahalka, M. 143

Zeitlmann, L. 33

Zutter, M.M. 167

VLA-4-Mediated Signaling

C. Morimoto[1,2], S. Iwata[1], and K. Tachibana[1]

1 Introduction . . . 1
2 Tyrosine Phosphorylation in T Cells Through Ligation of VLA-4 . . . 3
3 Identification of the Candidate Proteins . . . 4
4 Identification of a Tyrosine Phosphorylated, 70 kDa, pp125^{FAK}-Associated Protein as Paxillin . . . 5
5 Direct Association of pp125^{FAK} with Paxillin . . . 8
6 Identification of Paxillin-Binding Domain of pp125^{FAK} . . . 9
7 Identification of pp125^{FAK} Amino Acid Residues Essential for Paxillin Binding . . . 9
8 Biological Relevance Between Paxillin Binding and the FAT Domain of pp125^{FAK} . . . 11
9 Association of pp105 with the COOH-Terminal Domain of FAK . . . 12
10 Identification of pp105 as a Cas-Related Protein . . . 13
11 Differential Expression of 130 and 105 kDa Cas Proteins . . . 14
12 cDNA Cloning of pp105 . . . 15
13 Identification of the pp105-Binding Proteins . . . 16
14 Concluding Remarks . . . 17
References . . . 20

1 Introduction

The VLA (very late antigens) constitute the β1 subfamily of integrin adhesion receptors defined by at least nine α-chains that share a noncovalently linked common β-chain, termed β1 (CD29) (Hynes 1992; Schwartz 1993). The VLA mainly function as cell surface receptors mediating cell-to-cell and cell-to-extracellular matrix (ECM) adhesive interactions. They constitute a major class of adhesive receptors expressed by T cells. On resting CD4^{+} T cells, the VLA/CD29 antigens are preferentially expressed on the CD45RO^{+}CD45RA^{-} helper/inducer (memory)

[1]Division of Tumor Immunology, Dana-Farber Cancer Institute, 44 Binney Street, Boston, Massachusetts 02115, USA
[2]Department of Clinical Immunology and AIDS Research Center, Institute of Medical Science, University of Tokyo, 4-6-1, Shirokanedai, Minato-ku, Tokyo, 108, Japan

subset (Morimoto et al. 1985). VLA molecules are thought to play a major role in the interaction between these helper cells and the surrounding ECM or aid their migration into tissues (Shimizu and Shaw 1991). Apart from a role in cell adhesion, recent studies have clearly shown the VLA receptors to transduce signals in a wide variety of cells, including T lymphocytes (Schwartz et al. 1991; Shattil and Brugge 1991; Sultan et al. 1991). For example, several laboratories, including ours, have shown that the binding of T cells with ECM through VLA-β1-integrins provides costimulatory signals for T cell proliferation (Matsuyama et al. 1989; Nojima et al. 1990; Yamada et al. 1991a, b; Ennis et al. 1993). VLA-β1-integrins are also reported to be involved in T or B cell differentiation through interaction with fibronectin (FN), expressed on stromal cells in thymus or bone marrow, respectively (Salomon et al. 1994; Williams et al. 1991).

One of the earliest events to occur upon T cell activation is an increase in protein tyrosine phosphorylation, and this was shown to be essential for subsequent T cell proliferation (Gupta et al. 1994). Since ligation of VLA-β1-integrins provides a costimulatory signal to the T cell receptor (TCR)/CD3 complex, we analyzed the effect of engagement of VLA-β1 on protein tyrosine phosphorylation. We showed that either adherence to the CS-1 domain of FN (see below) or monoclonal antibody (mAb) cross-linking of VLA-β1 or VLA-4 rapidly stimulated tyrosine phosphorylation of cellular proteins, including a 105 kDa protein (pp105) in human H9 T-lymphoblastic cells or peripheral resting T cells. In B cells, ligation of VLA-4 induces tyrosine phosphorylation of a 105–125 kDa group of proteins (Nojima et al. 1992). Also, the interaction of VLA-4 with FN generates focal adhesions. Various cytoskeletal elements and a protein tyrosine kinase (PTK), $pp125^{FAK}$ (focal adhesion kinase), have been reported to sequester to the focal adhesive loci (Schaller et al. 1992). A recent study also identified highly tyrosine phosphorylated proteins of 120–130 kDa upon engagement of VLA-β1-integrins in different types of cells, including mouse fibroblasts (Guan et al. 1991) and a human epidermal cancer cell line (Kornberg et al. 1991), subsequently identified as various tyrosine phosphorylated forms of $pp125^{FAK}$ (Schaller et al. 1992). Our previous study demonstrated that pp105 was distinct from $pp125^{FAK}$ and that both pp105 and $pp125^{FAK}$ were tyrosine phosphorylated by the ligation of VLA in an H9 T cell (Nojima et al. 1995). Given the costimulation provided by VLA- β1-integrins and the putative role of $pp125^{FAK}$ in cell growth and differentiation, the above findings strongly suggest that pp105/$pp125^{FAK}$ tyrosine phosphorylation may play a crucial role in VLA-β1-integrin-mediated signaling events in T cell activation.

Although most members of the VLA family are involved in cell-ECM interactions, only VLA-4 has been conclusively shown to participate in both cell-ECM and cell-cell adhesive interactions. In particular, VLA-4 has been demonstrated to serve as a receptor for an Arg-Gly-Asp-independent site of plasma FN, namely CS-1, as well as for the cell surface molecule VCAM-1, a member of the Ig superfamily expressed on cytokine-activated endothelial cells. Moreover, VLA-4 mediates intercellular adhesion of leukocytes based on the ability of specific anti-VLA-4 mAbs to trigger homotypic cell aggregation through an LFA-1/ICAM-1-independent mechanism. Furthermore, accumulating evidence suggests that VLA-4 integrin-dependent adhesion pathways are critical intervention points in several inflammatory and autoimmune pathologies. In this chapter, we will focus on VLA-4-mediated signal transduction, especially results based on our previous and recent studies.

2 Tyrosine Phosphorylation in T Cells Through Ligation of VLA-4

Recently, VLA-4 has been found to mediate intercellular adhesion of leukocytes on the basis of the ability of specific anti-VLA-4 mAbs to trigger homotypic cell aggregation through an LFA-1/ICAM-1-independent mechanism (Sanchez et al. 1993). In this regard, several distinct VLA-4 adhesion functions have been reported. VLA-4-mediated homotypic cell aggregation, cell attachment to the CS-1 domain of FN, and adhesion to VCAM-1 can each be independently inhibited. This indicated that there are three distinct functions mediated by VLA-4. In addition, since some VLA-4 mAbs are reported to block homotypic cell aggregation but do not trigger it, we can consider these two aspects of the cell as functionally distinct. We showed that VLA-4 mAb against epitope B2 induces the strongest costimulation through the CD3 pathway. VLA-4 mAb against A and B1 epitopes can induce modest T cell costimulation, whereas VLA-4 mAb against the C epitope induces only minimal T cell costimulation. Thus, taken all together, the VLA-4 molecule is involved in a total of five distinct functions.

The fact that the anti-epitope B2 mAb was able to inhibit cell binding to both CS-1 and VCAM-1 indicated that VLA-4-mediated costimulation can be triggered through an epitope overlapping with the binding sites of CS-1 and VCAM-1. This is in agreement with previous observations, since both CS-1 and VCAM-1 can induce TCR/CD3-mediated costimulation. In our earlier study (Nojima et al. 1992), we demonstrated that liquid cross-linking of VLA-4 by one of the anti-VLA-4 mAbs, 8F2, and by an anti-CD29 mAb, 4B4, can induce tyrosine phosphorylation of a 105 kDa protein. Here, we analyzed the differences in the nature of protein tyrosine phosphorylation induced by antibody ligation of anti-VLA-4 epitope C (8F2), which appears to have no role in either cell aggregation or cell proliferation. In agreement with our previous report (Nojima et al. 1992), tyrosine phosphorylation of a 105 kDa protein in peripheral T cells was induced by the liquid phase cross-linking of VLA-4 using 4B4 as well as 8F2 and 3G6 after 10 min of stimulation (peak phosphorylation at 10–15 min). By contrast, after 10 min the solid phase cross-linking of VLA-4 using 4B4 and 3G6 induced various tyrosine phosphorylated proteins migrating at 140, 120, 110–105, 80–70, 60–55, 50, and 45 kDa on an SDS-PAGE (Sato et al. 1995). As was the case with liquid phase cross-linking of VLA-4, solid phase cross-linking using the high concentration of 8F2 (20 μg/ml) induced several weakly tyrosine phosphorylated proteins, such as 140, 120, and 80–70 kDa, but the main band was at 105 kDa. It should be noted that some tyrosine phosphorylated proteins (50 and 45 kDa proteins) were not so strong at 10 min.

As the difference in the tyrosine phosphorylation level was observed in the solid phase cross-linking between 3G6 and 8F2 stimulation, it was important to define the different signaling events (tyrosine phosphorylation) associated with VLA-4-mediated T cell costimulation. Therefore, we proceeded to compare the tyrosine phosphorylation induced by solid phase cross-linking of VLA-4 using 3G6 to that induced by its natural ligand, glutathione-S-transferase (GST)-CS-1. Solid phase cross-linking using and GST-CS-1 induced the same tyrosine phosphorylated proteins. In addition, the tyrosine phosphorylation induced by GST-CS-1 was almost completely inhibited by pretreating peripheral T cells with mAb 4B4 or 3G6.

To further determine the relationship between the VLA-4 epitope and tyrosine phosphorylation, we compared the tyrosine phosphorylation induced by solid phase cross-linking using epitope-specific anti-VLA-4 mAb. Three immobilized mAbs against B2 induced the same protein tyrosine phosphorylation as well as 4B4 (anti-β1 mAb), and immobilized mAb against A and B1 were able to induce the same protein tyrosine phosphorylation at 30 min of incubation. Three immobilized antibodies against epitope C could induce some tyrosine phosphorylation, but the intensity was weak compared with that of A or B epitope mAb. Since T cell costimulation could not be induced by C epitope mAb, these findings strongly suggest that the above tyrosine phosphorylated proteins induced by β2-epitope mAb may be closely involved in VLA-4-mediated T cell costimulatory events.

3 Identification of the Candidate Proteins

We next identified each protein candidate from the above bands by immunoprecipitation using antibody that recognizes the candidate protein. For this purpose, lysates from the cells that had been stimulated with solid phase cross-linked VLA-4 using mAb (8F2 and 3G6) were immunoprecipitated with the candidate mAb. The known substrate proteins, $pp125^{FAK}$, paxillin, Fyn, and Lck were clearly tyrosine phosphorylated in cells stimulated by solid phase cross-linking using 3G6, but not 8F2. This result indicated that the pp120 protein was $pp125^{FAK}$, pp70 and pp50 proteins were paxillin (two bands at 70 and 50 kDa were detected by anti-paxillin mAb), and pp60–55 proteins were $p59^{fyn}$ and $p56^{lck}$. Regarding the 140 kDa protein, similar immunoprecipitation studies demonstrated that antibody to IRS-1, JAK1, JAK2, Tyk2, Stat2, and β1-integrins, which recognize 130–160 kDa proteins (Sun et al. 1991; Horak et al. 1991; Silvennoinen et al. 1993; Firmbach et al. 1990; Johansson et al. 1994) were unable to reprecipitate phosphoprotein. In contrast, the antibody against phospholipase C (PLC)γ1 showed that p140 was actually PLCγ. In the case of pp45, a tyrosine-phosphorylated mitogen-activated protein kinase (MAPK) (ERK1/2) was detected on a solid phase anti-CD3-stimulated, or a solid phase 3G6-stimulated, sample but not on a BSA an 8F2 sample

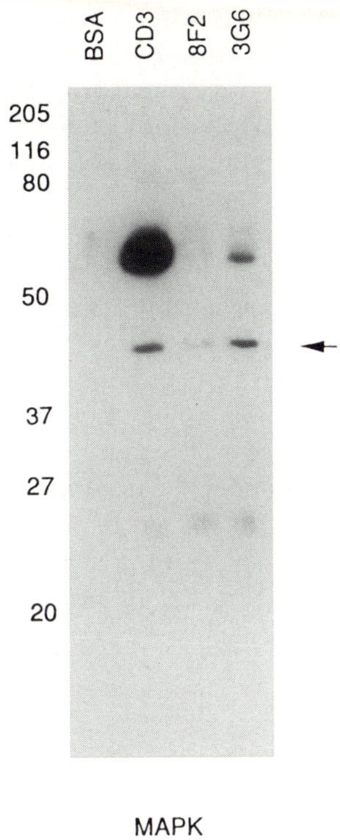

Fig. 1. Solid phase cross-linking of VLA-4 with 3G6 induces tyrosine phosphorylation of mitogen-activated protein kinase (MAPK). Resting T cells were incubated on plates coated with BSA (1%, W/V), anti-CD3, 8F2, or 3G6 (5 μg/ml of each) for 30 min and lysed. Lysates were immunoprecipitated with anti-phosphotyrosine beads, and immunoprecipitates were analyzed by anti-MAPK antibody immunoblotting using Enhanced Chemiluminescense (ECL, distributed by Amersham Life Science Inc. UK). The position of MAPK is indicated by the *arrowhead*

(Fig. 1) (SATO et al. 1995). Regarding the 55 kDa protein, which was detected after anti-CD3 stimulation or anti-VLA-4 stimulation, we have not succeeded in its identification. Therefore, it is not clear whether the anti-MAPK antibody cross-reacted with this protein, or if the 55 kDa protein also belongs to the MAPK family. In the cases of pp105 and pp80, the structure and nature of these proteins have not been clarified at the moment. Because solid phase cross-linking of VLA-4 by B2 epitope-specific antibody induced T cell costimulation most strongly through the CD3/TCR pathway, the above results strongly suggested that the above-mentioned tyrosine phosphorylated proteins may play an important role in VLA-4-mediated T cell costimulatory signaling events.

4 Identification of a Tyrosine Phosphorylated, 70 kDa, pp125^{FAK}-Associated Protein as Paxillin

Focal adhesion kinase, pp125^{FAK}, colocalizes with VLA-β1-integrins at focal adhesions, where cells attach to the ECM. The complex architecture of cellular focal adhesions depends on numerous protein-protein interactions between focal adhesion components. However, the mechanisms and cellular components regulating various aspects of signaling induced by VLA-β1-integrins are currently unclear. Because VLA-β1-integrins lack intrinsic PTK activity, the above protein tyrosine phosphorylations strongly suggest the presence of PTK(s), which is functionally linked to VLA-β1-integrins. $pp125^{FAK}$ is a prime candidate for such PTKs because it is colocalized with VLA-β1-integrins at focal adhesions.

It is now established that one of the major substrates for integrin-mediated tyrosine phosphorylation is $pp125^{FAK}$, a 125 kDa cytoplasmic PTK. To determine the relationship between $pp125^{FAK}$ and other tyrosine phosphorylated proteins, we attempted to define $pp125^{FAK}$-associated molecules. For this purpose, lysates of H9 cells or T-47D cells were precipitated with the GST fusion protein containing $pp125^{FAK}$ COOH-terminal domain and were analyzed by immunoblotting with anti-phosphotyrosine mAb. As shown in Fig. 2, tyrosine phosphorylated $pp125^{FAK}$ was well precipitated with GST-paxillin fusion protein in nonlymphoid adherent cells such as T-47D, a human breast cancer cell line (Tachibana et al. 1995). Furthermore, a tyrosine phosphorylated 70 kDa protein was shown to precipitate with GST-FAK fusion protein. In addition to adherent cells, the 70 kDa protein,

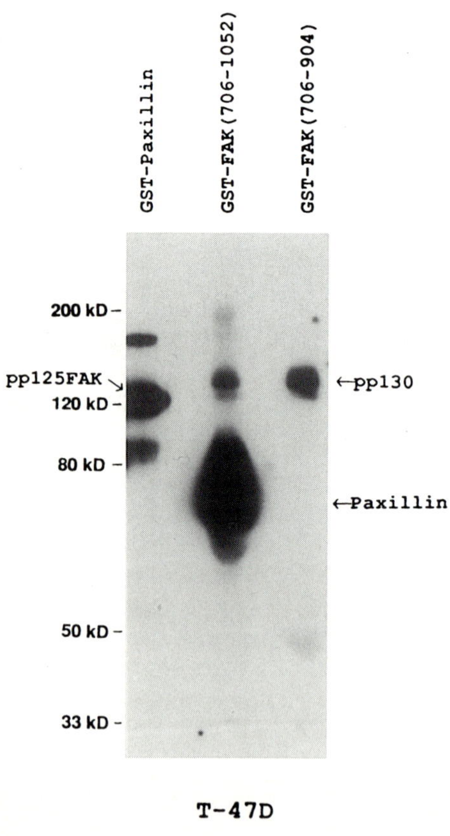

Fig. 2. Identification of a tyrosine phosphorylated 70 kDa $pp125^{FAK}$-associated protein (FAK, focal adhesion kinase). T-47D cell lysates were precipitated with glutathione-S-transferase (GST)-paxillin fusion protein (*lane 1*) and GST-$pp125^{FAK}$ fusion proteins (*lanes 2, 3*). The precipitates were subjected to SDS-PAGE and analyzed by immunoblotting with ^{125}I-labeled anti-phosphotyrosine monoclonal antibody (4G10)

which was precipitated by the GST-FAK fusion protein from the lysate of FN-stimulated H9 cells, was also detected by immunoblotting with antiphosphotyrosine mAb (Fig. 3B). This 70 kDa protein was not detected by antiphosphotyrosine mAb blotting in the absence of FN stimulation (Fig. 3A). These results suggest the following: (a) the 70 kDa protein is tyrosine phosphorylated by FN stimulation in H9 cells; (b) this 70 kDa protein (pp70) binds to FAK protein, although it is not clear whether tyrosine phosphorylation of this protein is required for binding to $pp125^{FAK}$. Tyrosine phosphorylation in the binding site of $pp125^{FAK}$ is not necessary for pp70 binding because the GST-FAK fusion protein, which was not phosphorylated on tyrosine residues, bound to pp70.

To identify the pp70 protein, pp70 was precipitated with the GST-FAK fusion protein, fractionated by SDS-PAGE, and analyzed by immunoblotting with specific

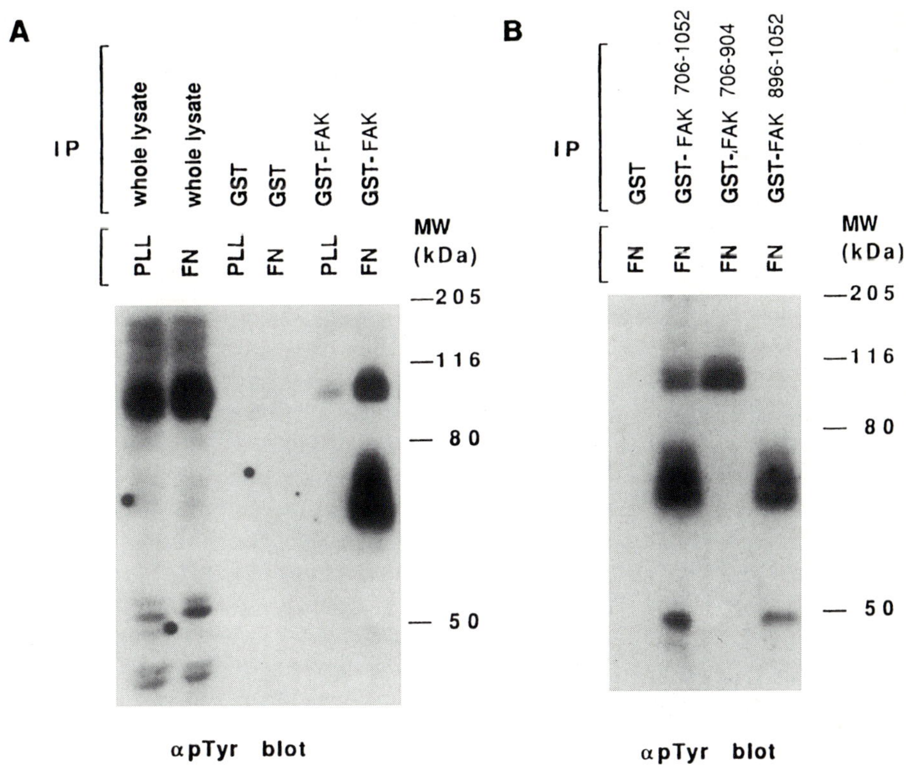

Fig. 3A, B. Binding of pp105 to the CT domain of focal adhesion kinase (FAK). **A** H9 cells were incubated in poly-L-lysine (PLL)- or fibronection (FN)-coated plates for 30 min (*lanes 1, 3, 5*, and *2, 4, 6*, respectively). After lysis, cellular lysates were precipitated by glutathione beads that were conjugated to glutathione-S- transferase (GST) or GST-FAK COOH-terminal domain (CT) fusion protein (containing FAK residues 706–1052, GST-CT beads). The whole lysate and the immunoprecipitates were analyzed by immunoblotting with anti-phosphotyrosine monoclonal antibody (α-pTyr). **B** H9 cell lysates were precipitated with GST-FAK fusion protein-conjugated beads. Precipitates were analyzed by immunoblotting with α-pTyr. For discussion, see text antibodies against candidate proteins for pp70. We found that paxillin was precipitated with GST-FAK fusion protein from HPB-ALL (a human T lymphoblastic cell line) lysates with or without FN stimulation, as well as from T-47D cell lysate, and it migrated with a mobility similar to that of pp70, suggesting that pp70 was identical to the tyrosine phosphorylated paxillin. To confirm that pp70 was paxillin, pp70 was precipitated by the GST-FAK fusion protein from T-47D cell lysate, solubilized by boiling in buffer containing 1% SDS, reprecipitated with anti-phosphotyrosine mAb or anti-paxillin mAb, and analyzed by immunoblotting with antipaxillin mAb and with antiphosphotyrosine mAb. Tyrosine phosphorylated pp70, which was precipitated with the GST-FAK fusion protein, was detected by anti-paxillin mAb, and tyrosine phosphorylated paxillin migrated with exactly the same mobility as pp70. These results confirm that pp70 is paxillin. Further studies showed that nontyrosine phosphorylated paxillin was precipitated with the GST-FAK fusion protein as well as pp70, tyrosine phosphorylated paxillin. These results strongly suggest that tyrosine phosphorylation of paxillin is not required for binding to $pp125^{FAK}$.

5 Direct Association of $pp125^{FAK}$ with Paxillin

Although the association of the GST-FAK fusion protein with paxillin was demonstrated, the following points remained unclear: (a) Is endogenous $pp125^{FAK}$

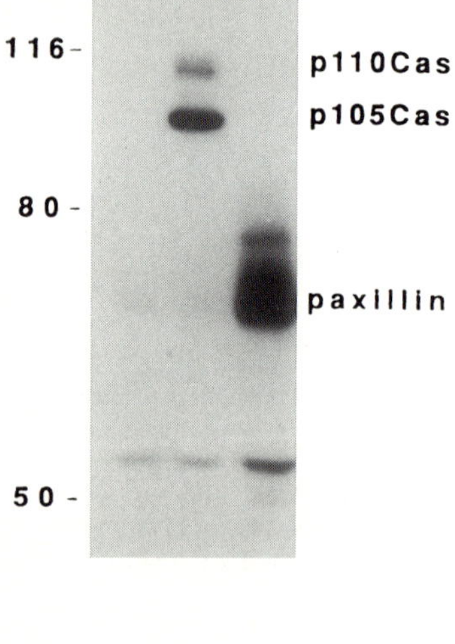

Fig. 4. Identification of the focal adhesion kinase (FAK)-binding proteins as paxillin and pp105/110 Cas-related proteins. H9 cell lysates were immunoprecipitated without first antibody (*lane 1*), with anti-Cas (Crk-asociated substrate) monoclonal antibody (mAb) (*lane 2*), and with anti-paxillin mAb (*lane 3*). After fractionation by SDS-PAGE and electrotransfer, immunoprecipitates were denatured and renatured, followed by overlay with ^{32}P-labeled glutathione-S-transferase (GST)-FAK fusion protein containing FAK residues 706–1052

associated with paxillin? (b) Is paxillin directly associated with pp125^{FAK} or associated indirectly via binding to other proteins?

To determine whether the association between pp125^{FAK} and paxillin was direct or indirect, we performed an overlay assay using ^{125}I-labeled GST-FAK fusion protein as a probe. Immunoprecipitated paxillin was analyzed by overlay assay with ^{125}I-labeled GST-FAK fusion protein. As shown in Fig. 4, a protein of 70 kDa was detected on the lanes of precipitates with anti-paxillin mAb, whereas no protein was detected using control precipitations (Tachibana et al. 1995). This protein had the same mobility as paxillin as shown by reprobing of the membrane with anti-paxillin mAb. These results demonstrate a direct association between pp125^{FAK} and paxillin.

6 Identification of Paxillin-Binding Domain of pp125^{FAK}

To determine the paxillin-binding domain of pp125^{FAK}, we developed several deletion mutants derived from the GST-FAK fusion protein. HPB-ALL cell lysates were incubated with these deletion mutant proteins, precipitated, and analyzed by immunoblotting with anti-paxillin mAb. Our results showed that pp125^{FAK} residues 919–1042 are sufficient for the association with paxillin. Deletion mutants containing FAK residues 923–1052 or 896–1039 did not precipitate paxillin, suggesting that the amino acid residues of pp125^{FAK}, which are critical for paxillin binding, are located close to both ends of the paxillin-binding domain, residues 919–1042 (Table 1) (Tachibana et al. 1995). This paxillin-binding domain of human pp125^{FAK} overlaps with the FAT (focal adhesion targeting) domain of chick pp125^{FAK}. Hildebrand et al. (1993) reported that chick pp125^{FAK} mutants with a deletion that corresponds to human pp125^{FAK} residues 861–967 or 969–1016 resulted in the loss of FAT. Since these two mutants have deletions in the paxillin-binding domain, these mutants are expected to lack paxillin-binding activity, suggesting the relevance between paxillin-binding activity and FAT of pp125^{FAK}.

7 Identification of pp125^{FAK} Amino Acid Residues Essential for Paxillin Binding

Paxillin was first identified as a vinculin-binding protein. Vinculin, a 116 kDa cytoskeleton protein, is also localized at focal adhesions, and the paxillin-binding domain of vinculin was identified recently (Wood et al. 1994). Chick vinculin residues 881–1000 are sufficient for paxillin binding, and vinculin residues 979–1000 are critical for paxillin binding. Comparison of paxillin-binding domains between pp125^{FAK} and vinculin revealed several conserved amino acid residues. As shown

Table 1. Summary of the functions of GST-FAK fusion proteins

Deletion mutants	10G2[a]	PB[b]	FAT[c]	Substitution mutants	10G2	PB	FAT
706–1052	+	+	+	923 K → E	+	+	+
706–841	–	–	ND[d]	928 V → G	±	–	–
706–904	–	–	–	929 T → A	+	+	ND
706–997	–	–	ND	929 T → S	+	+	ND
706–1038	+	–	–	931 L → R	±	–	–
896–1005	–	–	ND	933 K → E	+	+	ND
896–1015	–	–	ND	935 V → A	+	–	ND
896–1027	–	–	ND	1033 N → D	+	+	ND
896–1038	+	–	–	1034 L → S	+	–	–
896–1039	+	–	–	1035 L → A	+	–	ND
896–1042	+	+	ND	1036 D → H	+	+	+
896–1047	+	+	ND	1037 V → D	+	–	ND
896–1052	+	+	+	1039 D → A	+	–	ND
904–1052	+	+	+	1040 Q → E	+	+	ND
904–997	–	–	ND	1040 Q → G	+	+	ND
919–1052	+	+	+	1040 Q → K	+	+	ND
923–1052	+	–	ND	1042 R → G	+	–	ND
928–1052	+	–	–	1043 L → R	+	+	+
939–1052	+	–	–				
967–1052	±	–	ND				

Paxillin-binding activity, focal adhesion-targeting activity (FAT), and 10G2 mAb-binding activity of each GST-FAK fusion proteins is summarized. Paxillin-binding activity represents the activity needed to precipitate paxillin from cellular lysates. Focal adhesion-targeting activity was determined by immunohistochemical analysis of the microinjected fusion protein. 10G2 mAb-binding activity was determined by immunoblotting with ^{125}I-labeled 10G2 mAb.
[a]10G2 mAb binding activity.
[b]Paxillin binding activity.
[c]Focal adhesion targeting activity.
[d]Not determined.

in Fig. 5, two subdomains located on both ends of the paxillin-binding domain of pp125^{FAK} have homology with the paxillin-binding domain of vinculin (Tachibana et al. 1995). pp125^{FAK} residues 919–935 and vinculin residues 952–968 share a sequence, R/K-X6-E/D-X-V-T-X-L-X3-V/L (paxillin-binding subdomain 1/PBS1). pp125^{FAK} residues 1034–1039 and vinculin residues 981–986 also share a sequence, L-L-X-V-D/E (paxillin-binding subdomain 2/PBS2). To determine the role of these homologous amino acid residues in the interaction with paxillin, we generated substitution mutants within pp125^{FAK} residues 896–1052 and performed paxillin-binding analysis with these mutant proteins. Substitutions of residues Val-928 to Gly, Leu-931 to Arg, and Val-935 to Ala in PBS1 resulted in a significant decrease in paxillin-binding activity, whereas substitutions of Lys-923 to Glu, Thr-929 to Ala or Ser, and Lys-933 to Glu had no effect on paxillin binding. In PBS2, substitutions of Leu-1034 to Ser, Leu-1035 to Ala, Val-1037 to Asp, Glu-1039 to Ala, and Arg-1042 to Gly significantly decreased paxillin-binding activity, whereas Asn-1033 to Asp, Glu-1036 to His, Gln-1040 to Glu, Gly, or Lys, and Leu-1043 to Arg did not. These results clearly indicate that conserved or homologous amino acids between pp125^{FAK} and vinculin are essential for paxillin binding.

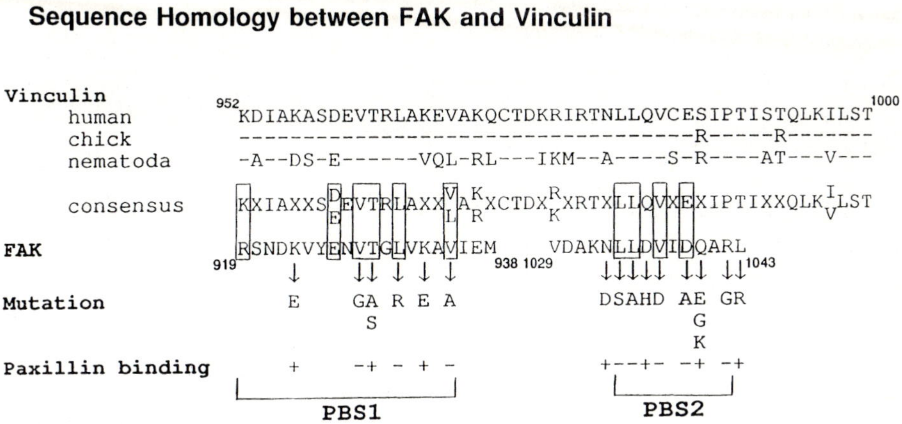

Fig. 5. Sequence homology between pp125^{FAK} (FAK, focal adhesion kinase) and vinculin. Amino acid sequences of the paxillin-binding domains of vinculin and pp125^{FAK}. Amino acid sequences of the paxillin-binding domains of human, mouse, and chick pp125^{FAK} are conserved. Conserved amino acids between pp125^{FAK} and vinculin are *boxed*. Substituted FAK residues used in the following analysis and paxillin-binding activity of each mutant are shown below FAK sequence

8 Biological Relevance Between Paxillin Binding and the FAT Domain of pp125^{FAK}

Given that the paxillin-binding domain and the FAT domain of pp125^{FAK} overlap each other, one of the putative biological functions of pp125^{FAK} binding to paxillin would be the recruitment of pp125^{FAK} to focal adhesions. To elucidate this possibility, we performed immunohistochemical analysis combined with microinjection of GST-FAK fusion proteins. The results clearly depict the relevance between paxillin-binding activity and FAT activity of pp125^{FAK}, indicating that pp125^{FAK} localizes to focal adhesions by its direct interaction with paxillin (Tachibana et al. 1995). Deletion and substitution mutants with paxillin-binding activity are localized to focal adhesions, whereas mutants without paxillin-binding activity are not localized to focal adhesions. These findings indicate that pp125^{FAK} is localized to focal adhesions by association with paxillin. Focal adhesions, where cells attach to substrata via integrin-ECM binding, are also composed of cytoskeletal proteins such as talin, vinculin, paxillin, FRNK (FAK-related nonkinase), and pp125^{FAK} (Burridge et al. 1988, 1992). Among these proteins, vinculin, which has a conserved paxillin-binding domain with pp125^{FAK}, appears to be localized to focal adhesions by association with paxillin (Wood et al. 1994), and talin is associated with vinculin (Burridge and Mangeat 1984). Our findings strongly suggest that pp125^{FAK} and other proteins are localized to focal adhesions and linked to β1-integrins via interaction with paxillin. β1-integrin-ligand ligation induces FAT, and tyrosine phosphorylation of pp125^{FAK} then induces recruitment of various signaling molecules to tyrosine phosphorylated pp125^{FAK}, which presumably results in β1-integrin-mediated cell adhesion, cytoskeleton organization, and cell proliferation.

9 Association of pp105 with the COOH-Terminal Domain of FAK

Although we identified several tyrosine phosphorylated proteins stimulated by the ligation of VLA-β1-integrins in T cells, a 105 kDa tyrosine phosphorylated protein (pp105) had not yet been identified. pp105 is a protein which we first identified in T lymphoblastoid H9 cells as well as peripheral T cells and is tyrosine phosphorylated by the engagement of VLA-4. In our previous study (Nojima et al. 1995), we demonstrated that pp105 and pp125^{FAK} were tyrosine phosphorylated by VLA-β1-integrin stimulation via similar kinetics in H9 cells, although pp105 is a distinct molecule from pp125^{FAK}. Because pp125^{FAK} is an essential tyrosine kinase for VLA-β1-integrin-mediated protein tyrosine phosphorylation, we attempted to define the relationship between pp125^{FAK} and pp105. For this purpose, H9 cells were incubated with FN- or poly-L-lysine (PLL)-coated plates before cell lysis. H9 cell lysates were precipitated with a GST fusion protein of the pp125^{FAK} COOH-terminal domain (residues 706–1052, designated GST-CT) and analyzed by immunoblotting with anti-phosphotyrosine mAb (anti-pTyr). As shown in Fig. 3A, a tyrosine phosphorylated 105 kDa protein precipitated on beads conjugated with GST-CT from FN-stimulated cell lysate, whereas this protein did not precipitate on GST-conjugated beads. This tyrosine phosphorylated protein migrated at the same position as pp105 in FN-incubated cell lysate and was detected only minimally in PLL-incubated cell lysate (Minegishi et al. 1996). These results strongly suggest that pp105 binds to the CT of pp125^{FAK}. Paxillin was also precipitated from H9 cell lysate and was detected by anti-pTyr as a 70 kDa band, as we reported in HPB-ALL and T-47D cells (Tachibana et al. 1995). To determine whether pp105 binds to FAK or paxillin, H9 cell lysates were precipitated with deletion mutants of GST-CT. As shown in Fig. 3B, pp105 was precipitated with GST-FAK (residues 706–904), but not with GST-FAK (residues 896–1052). Conversely, paxillin was precipitated with GST-FAK (residues 896–1052), but not with GST-FAK (residues 706–904). These results demonstrate that pp105 binds to the FAK sequence that contains amino acid residues 706–904. This pp105 binding domain of FAK is distinct from the paxillin-binding domain of FAK, indicating that pp105/FAK binding is not mediated via paxillin. To further characterize pp105-FAK binding, pp105 was precipitated from H9 cell lysates with anti-Cas (Crk-associated substrate) mAb and analyzed by the overlay assay with ^{125}I-labeled GST-FAK fusion protein (GST-CT). As shown in Fig. 4, 105 and 110 kDa proteins were detected with labeled GST-CT, indicating that both proteins directly bind to FAK.

10 Identification of pp105 as a Cas-Related Protein

pp105 was precipitated with the GST-FAK fusion protein from FN-stimulated H9 cell lysates and was detected by immunoblotting with anti-pTyr. By contrast, a 130 kDa tyrosine phosphorylated protein was precipitated with the GST-FAK fusion protein from human breast cancer-derived T-47D cell lysates. We identified this 130 kDa protein as p130 Cas using anti-Cas mAb (Transduction Laboratories, Lexington, Ky, USA). Recently, Polte and Hanks (1995) reported that p130 Cas bound to FAK by its SH3 domain. FAK residues 706–904, which were sufficient for pp105 binding, contained the reported p130 Cas-binding site. We attempted to determine if pp105 was reactive with anti-Cas mAb. As shown in Fig. 6, a 105 kDa Cas protein was precipitated by the GST-FAK fusion protein and detected by immunoblotting with anti-Cas mAb. This 105 kDa protein showed the same migration in SDS-PAGE as pp105, which was precipitated by the GST-FAK fusion protein and detected by immunoblotting with anti-pTyr. We further performed a second immunoprecipitation with anti-Cas mAb after precipitation with the GST-FAK fusion protein. As shown in Fig. 6, a 105 kDa Cas protein was precipitated first by the GST-FAK fusion protein and reprecipitated by anti-Cas mAb (anti-Cas

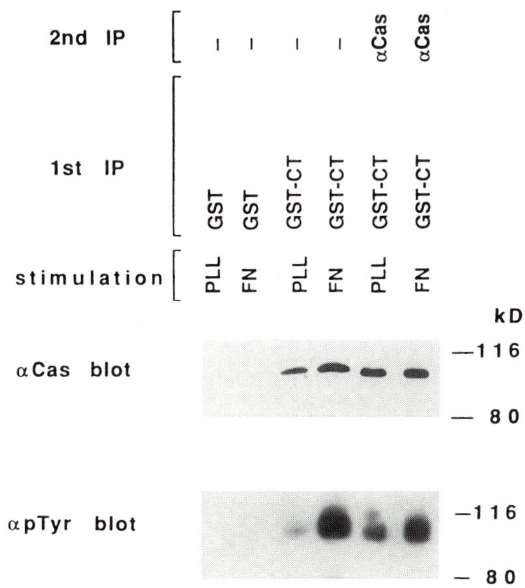

Fig. 6. Identification of pp105 as a Crk-associated substrate (Cas)-related protein. H9 cell lysates were prepared from cells that were incubated on poly-L-lysine (PLL)- or fibronectin (FN)-coated plates (*lanes 1, 3, 5* and *2, 4, 6*, respectively) for 30 min. Cellular lysates were precipitated with glutathione-S-transferase (GST) beads (*lanes 1, 2*) or GST-COOH terminal domain (CT) beads (*lanes 3–6*). After precipitation, associated molecules were solubilized and reprecipitated with anti-Cas (α-Cas) monoclonal antibody (mAb) (*lanes 5, 6*). Precipitates were analyzed by immunoblotting with anti-Cas mAb and rehybridized with anti-phosphotyrosine monoclonal antibody (α-pTyr).

blot). This 105 kDa protein showed increased tyrosine phosphorylation by FN stimulation (Fig. 6, anti-pTyr blot) (Minegishi et al. 1996).

To determine if this 105 kDa protein recognized by anti-Cas mAb is the major tyrosine phosphorylated protein among 105–110 kDa proteins that are phosphorylated by the ligation of VLA-β1-integrins, we performed immunodepletion analysis using anti-Cas mAb. A clear difference in the amount of tyrosine phosphorylated 105 kDa protein was demonstrated with or without immunodepletion with anti-Cas mAb both in the whole lysates and the anti-pTyr precipitates. These findings indicate that pp105, a 105 kDa protein tyrosine phosphorylated by VLA-β1-integrin stimulation, is a Cas-related protein.

11 Differential Expression of 130 and 105 kDa Cas Proteins

p130 Cas has been reported to migrate as a discrete species of 115 and 125 kDa in SDS-PAGE (designated by Sakai et al. 1994 as Cas-A and Cas-B, respectively) in rat fibroblast 3Y1 cells. However, a decrease in the size of Cas A and the simultaneous appearance of a broad 130 kDa Cas band (designated Cas-C) were observed in both v-Src- and v-Crk-transformed 3Y1 cells. Because phosphorylated tyrosine residues were found predominantly in Cas-C, the latter appeared to be a modified form of Cas-A or Cas-B as the result of tyrosine phosphorylation (Sakai et al. 1994).

We identified pp105 in H9 cells as a putative Cas-related protein. pp105 was originally identified as a 105 kDa protein that was tyrosine phosphorylated by the stimulation of VLA-β1-integrins. By immunoblotting with anti-pTyr, pp105 was detected predominantly in H9 cells as well as in peripheral T cells (Nojima et al. 1992), whereas pp105 was not detected well in other T cell lines such as Jurkat. To determine the distribution of pp105 and p130Cas, we examined the expression and the mobility in SDS-PAGE of Cas proteins in various cell lines. As shown in Fig. 7, using the same amount of whole extract from each cell line, Cas proteins with proteins similar to Cas-A and Cas-B from 3Y1 cells were detected in human breast cancer T-47D cells, although Cas proteins in T-47D cells showed slightly faster migration (Minegishi et al. 1996). Cas proteins of similar mobility to Cas-A protein of T-47D were also detected in human myelogenous cell lines HL-60 and K562. However, pp105 showed that a distinct mobility from Cas A and Cas-B was detected in the human myelogenous cell line U937 and in the human T lymphoblastic cell lines H9, HPB-ALL, Jurkat, and in the human B lymphoid cell line Raji. pp105 was also detected in human thymocytes and in human peripheral T and B cells. pp105 was significantly overexpressed in H9 cells, followed by peripheral T and B cells, thymocytes, and Raji cells. A 110 kDa Cas protein that migrated more slowly than pp105 but faster than Cas-A of T-47D cells was also detected in H9 cells, thymocytes, and peripheral T cells. These findings indicate that pp105 is a Cas-related protein that is preferentially expressed in lymphocytes.

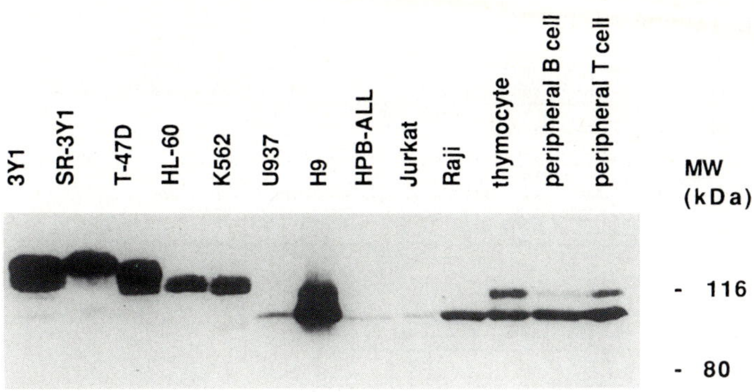

Fig. 7. Differential expression of 130 and 105 kDa Crk-associated substrate (Cas) proteins. Expression of Cas protein in each cell line was analyzed by immunoblotting with anti-Cas mAb (50 μg lysate/lane)

12 cDNA Cloning of pp105

To further determine the structure of pp105, we screened the λgt11 cDNA library derived from a human T lymphoblastoid cell line (Hut78) with anti-Cas mAb. Nucleotide sequences of three independent clones had homology with p130 Cas (MINEGISHI et al. 1996). These three clones were cDNAs of an identical transcript, and the nucleotide sequences contained an open reading frame of 834 amino acids. The deduced amino acid sequences of this transcript showed conserved motifs with p130 Cas, one SH3 domain in the NH_2-terminal region, and multiple putative binding sites for the SH2 domains (Fig. 8). Most of the SH2 binding motifs in the substrate domain are YXXP (YDXP), which are putative binding sites for Crk, Nck, and Abl SH2 domains (SONGYANG et al. 1993). Despite the conserved motifs, homology between p130 Cas and the deduced amino acid sequence of this cDNA is relatively low (78% in the SH3 domain, 32% in the substrate domain, 30% in the specific domain, and 32% in the CT). Homology with another Cas-related protein, Efs (ISHINO et al. 1995), is also relatively low. These results indicate that cDNA encodes a novel Cas-related protein.

To determine whether this cDNA of a novel Cas-related protein encodes pp105, the cDNA was inserted into an expression vector and transferred into Cos-1 cells. Cellular lysates from transfectants were analyzed by immunoprecipitation and immunoblotting. Subsequently, anti-Cas mAb-reactive peptides that migrated at 105 and 110 kDa in SDS-PAGE were detected with Cos-1 transfectant. pp105 from H9 cells comigrated with a 105 kDa peptide detected in the lysate from the transfectant, and a slightly slower-migrating Cas protein of 110 kDa in H9 cells comigrated with a 110 kDa peptide. This result strongly suggests that: (a) the Cas-related gene encodes pp105 and (b) a 110 kDa Cas protein detected in H9 cells is

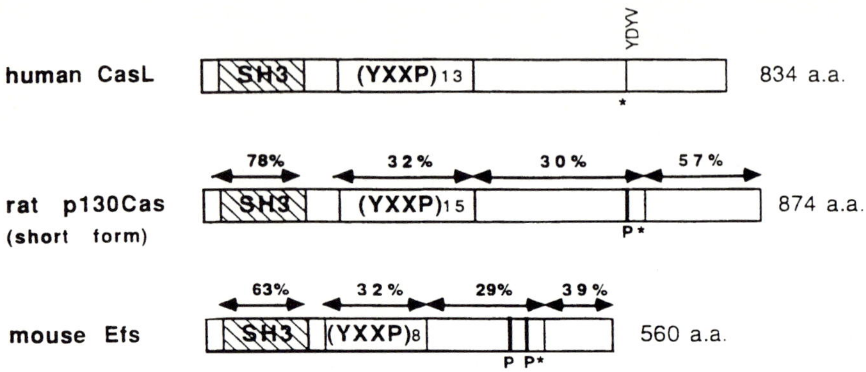

Fig. 8. Structure of pp105/Cas-L. Comparison of amino acid sequences among Crk-associated substrate (Cas) proteins. Amino acid sequence homologies between Cas-L and the other Cas protein in the SH3 domain, substrate domain (Cas-L residues 92–348), specific domain (Cas-L residues 349–628), and CT (Cas-L residues 629–834) are shown above each domain. YDYVHL motifs are shown by *asterisks* and *vertical lines*. Proline-rich sequences are shown by *P* and *thick vertical lines*

the protein product of the same transcript as pp105, despite a different mobility in SDS-PAGE (similar to that observed with p130 Cas-A and Cas-B).

p130 Cas was highly phosphorylated on tyrosine residues in v-Src- or v-Crk-expressing cells (SAKAI et al. 1994). Like p130 Cas, pp105 was also highly phosphorylated on tyrosine residues by cotransfection of Src, Lck, CrkI, or CrkII. Moreover, tyrosine phosphorylated pp105 binds to Crk proteins in vivo.

13 Identification of the pp105-Binding Proteins

We next attempted to define the proteins that were recruited to pp105 in a phosphorylated tyrosine residue-dependent manner. For this purpose, lysates from FN-stimulated H9 cells were precipitated with GST fusion proteins that contained src homology (SH)2 domains from various proteins and analyzed by immunoblotting with anti-Cas mAb and anti-pTyr. As a result, 110–105 kDa tyrosine phosphorylated proteins were precipitated with GST-fusion proteins of c-Abl, Crk, Csk, Grb2, Lck, Nck, and SHPTP2 SH2 domains (anti-pTyr blot). pp105 was precipitated by GST-AblSH2, GST-CrkSH2, and GST-NckSH2, whereas pp105 was weakly precipitated by GST-Lck SH2, GST-SHPTP2SH2, and GST-CskSH2 (anti-Cas Blot) (MINEGISHI et al. 1996). To determine whether the binding of pp105 to these GST fusion proteins was induced by VLA-β1-integrin stimulation, a similar analysis was performed using lysates of H9 cells that were incubated in PLL- or FN-coated plates. Subsequently, enhancement of pp105 precipitation was observed in the lanes of GST-SH2 domain fusion proteins of c-Abl, Crk, and Nck, whereas increased but slight amounts of pp105 were detected in the lanes of the GST-SH2

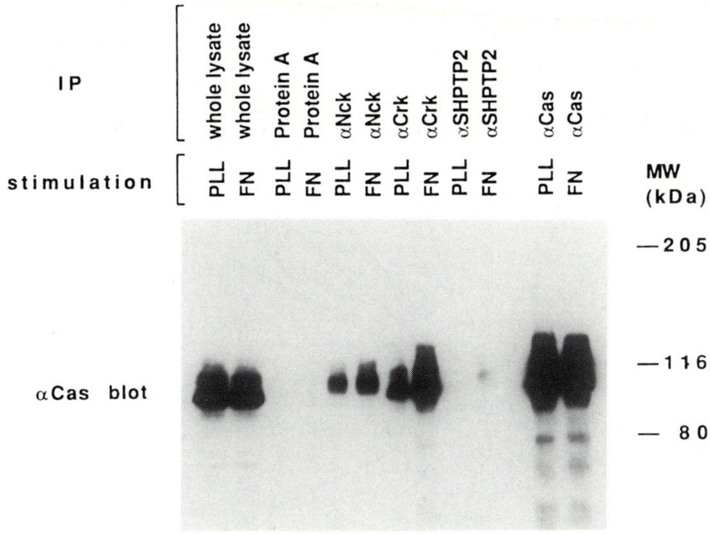

Fig. 9. Identification of the pp105-binding proteins. H9 cells were incubated in poly-L-lysine (PLL)- or fibronectin (FN)-coated plates (*lanes 1, 3, 5, 7, 9, 11* and lanes *2, 4, 6, 8, 10, 12*, respectively) for 30 min and lysed in 1% digitonin lysis buffer. Cellular lysates were immunoprecipitated without first antibody (*lanes 3, 4*) or with anti-Nck monoclonal antibody (mAb) (*lanes 5, 6*), anti-Crk mAb (*lanes 7, 8*), anti-SHPTP2 mAb (*lanes 9, 10*), or anti-Cas mAb (*lanes 11, 12*). Whole lysates (*lanes 1, 2*) and immunoprecipitates were analyzed by immunoblotting with anti-Cas mAb

domain fusion proteins of Lck and SHPTP2. These results indicated that tyrosine phosphorylated pp105 binds to SH2 domains of c-Abl, Crk, Lck, Nck, and SHPTP2 in vitro.

To further determine if these pp105-binding proteins bind to pp105 in vivo, co-immunoprecipitation analysis of pp105 with these proteins was performed. As shown in Fig. 9, pp105 was coprecipitated with Crk and Nck, whereas pp105 was weakly coprecipitated with SHPTP2. Unlike pp125^{FAK}-pp105 binding, pp105 that was coprecipitated with Crk, Nck, or SHPTP2 was increased by VLA-β1-integrin stimulation with FN. These results indicate that VLA-β1-integrin stimulation leads to the recruitment of various proteins, including Crk, Nck, and SHPTP2, to the tyrosine phosphorylated pp105, in addition to stimulation-independent association with pp125^{FAK}. These protein-protein interactions further suggest the putative function of pp105 in the VLA-β1-integrin-mediated signaling pathways.

14 Concluding Remarks

Initially, our studies attempted to elucidate the mechanism of the costimulatory nature of integrin engagement to TCR-mediated cell signaling in T lymphocytes (Matsuyama et al. 1989; Nojima et al. 1990, 1992). Subsequently, we found that a

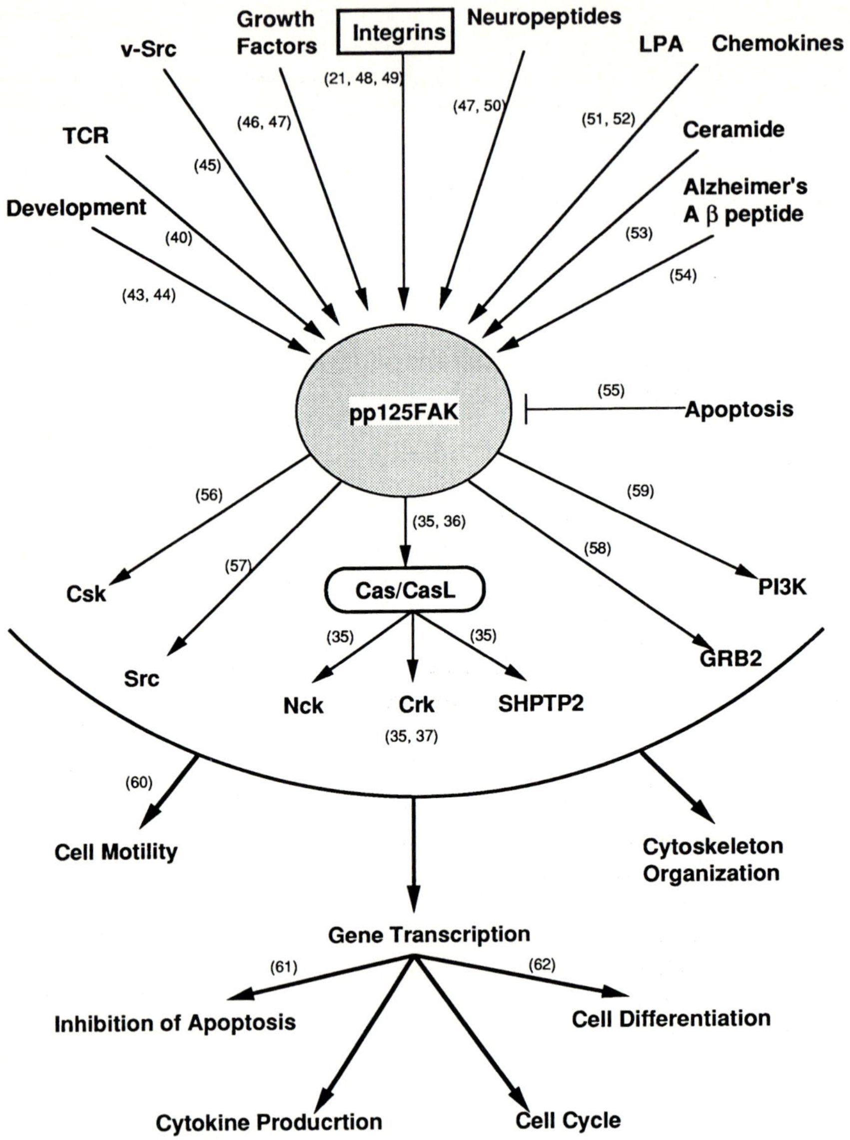

Fig. 10. Integrin-mediated signal transduction pathways. Lines with *arrowhead* mean some interactions such as binding, activation, etc. The line without the *arrowhead* suggests inhibition or degradation of $pp125^{FAK}$. Each *number* beside the line indicates the corresponding reference: (1) Polte et al. 1994; (2) Zhang et al. 1995; (3) Maquire et al. 1995; (4) Xing et al. 1994; (5) Hatai et al. 1994; (6) Rozengurt 1994a, b; (7) Nojima et al. 1995; (8) Schaller et al. 1994; (9) Lipfert et al. 1992; (10) Zachary et al. 1992; (11) Seufferlein and Rozengurt 1994a; (12) Bacon et al. 1996; (13) Seufferlein and Rozengurt 1994b; (14) Zhang et al. 1994; (15) Crouch et al. 1996; (16) Bachelot et al. 1996; (17) Schlaepfer et al. 1994; (18) Minegishi et al. 1996; (19) Polte and Hanks 1995; (20) Sakai et al. 1994; (21) Schaller et al. 1992; (22) Sabe et al. 1994; (23) Ilic et al. 1995; (24) Frisch et al. 1996; (25) Furuta et al. 1995

distinct set of proteins was phosphorylated on their tyrosine residues upon engagement of VLA-β1-integrin (Sato et al. 1995). Among those proteins, PLCγ, $p59^{fyn}$, $p56^{lck}$, and ERK1/2 are supposed to participate in TCR-mediated signaling pathways. Integrin-mediated tyrosine phosphorylation of those proteins may result in the augumentation or sustenance of TCR-mediated signal. As schematized in Fig. 10, a variety of stimuli beside the engagement of integrins are reported to induce tyrosine phosphorylation of $pp125^{FAK}$. It was shown that ligation of TCR causes, in turn, tyrosine phosphorylation of $pp125^{FAK}$ (Maguire et al. 1995), indicating that $pp125^{FAK}$ might be a key molecule involved in coordination of TCR- and integrin-mediated signals. These findings may provide an important clue regarding cross-talk between TCR- and integrin-mediated signaling pathways. Another major question concerns the missing link between binding of Crk to pp105/Cas-L and transcriptional regulation of various genes which occurs on T cell activation. Our results (Tachibana et al. 1995) suggest that phosphorylation-dependent binding of Crk, Nck, and SHPTP2 may be the downstream events of integrin-mediated tyrosine phosphorylation of pp105/Cas-L. However, the biological outcome of those events still remains to be elucidated. The most immediate goals are to determine the biological relevance of tyrosine phosphorylation of pp105/Cas-L and $pp125^{FAK}$. Since all the integrin molecules lack a putative enzymatic domain for signal transduction, PTKs that preferentially localize to focal adhesion sites, $pp125^{FAK}$ and Src are potential candiates for the molecules responsible for integrin-mediated signal transduction. Indeed, Src is reported to phosphorylate p130Cas. Is pp105/Cas-L the substrate of $pp125^{FAK}$? What is the subcellular localization of pp105/Cas-L? Although it is possible that pp105/Cas-L may be recruited to focal adhesion through the interaction with $pp125^{FAK}$, the precise manner of pp105/Cas-L distribution has yet to be determined. Further, our study revealed that $pp125^{FAK}$ is targeted to focal adhesion with paxillin in a phosphorylation-independent manner (Tachibana et al. 1995). Then what is the mechanism of the induction of tyrosine phosphorylation on those proteins such as $pp125^{FAK}$, pp105/Cas-L, and paxillin? The activation mechanism of the kinase activity of $pp125^{FAK}$ needs to be investigated. In the light of recent data on the regulation of integrin avidity by various small G proteins such as R-ras and a constitutive active mutant of ras (Zhang et al. 1996; Hughes et al. 1997), efforts should be made to determine the effect of those proteins on phosphorylation and/or activation of $pp125^{FAK}$ and pp105/Cas-L. Finally, these approaches will provide some insights into therapeutic intervention of integrin-mediated processes which are involved in a variety of diseases.

Acknowledgements. This work was in part supported by NIH Grants AR-33713 and AI-29530 and in part by the grants from the Ministry of Education, Science, Sports, and Culture and the Ministry of Health Welfare of the Japanese government.

References

Bachelot C, Rameh L, Parsons T, Cantley LC (1996) Association of phosphatidylinositol 3-kinase, via the SH2 domains of p85, with focal adhesion kinase in polyoma middle t-transformed fibroblasts. Biochim Biophys Acta 1311:45–52

Bacon KB, Szabo MC, Yssel H, Bolen JB, Schall TJ (1996) RANTES induces tyrosine kinase activity of stably complezed p125FAK and ZAP-70 in human T cells. J Exp Med 184:873–882

Burridge K, Mangeat P (1984) An interaction between vinculin and talin. Nature 308:744–746

Burridge K, Turner CE, Romer LH (1992) Tyrosine phosphorylation of paxillin and pp125FAK accompanies cell adhesion to extracellular matrix: a role in cytoskeletal assembly. J Cell Biol 119:893–903

Burridge K, Fath K, Kelly T, Nuckolls G, Turner C (1988) Focal adhesions: transmembrane junctions between the extracellular matrix and the cytoskeleton. Annu Rev Cell Biol 4:487–525

Crouch DH, Fincham VJ, Frame MC (1996) Targeted proteolysis of the focal adhesion kinase pp125 FAK during c-MYC-induced apoptosis is suppressed by integrin signalling. Oncogene 12:2689–2696

Ennis E, Isberg RR, Shimizu Y (1993) Very late antigen 4-dependent adhesion and costimulation of resting human T cells by the bacterial beta 1 integrin ligand invasin. J Exp Med 177:207–212

Firmbach KI, Byers M, Shows T, Dalla FR, Krolewski JJ (1990) tyk2, prototype of a novel class of nonreceptor tyrosine kinase genes. Oncogene 5:1329–1336

Frisch SM, Vuori K, Ruoslahti E, Chan HP (1996) Control of adhesion-dependent cell survival by focal adhesion kinase. J Cell Biol 134:793–799

Furuta Y, Ilic D, Kanazawa S, Takeda N, Yamamoto T, Aizawa S (1995) Mesodermal defect in late phase of gastrulation by a targeted mutation of focal adhesion kinase, FAK. Oncogene 11:1989–1995

Guan JL, Trevithick JE, Hynes RO (1991) Fibronectin/integrin interaction induces tyrosine phosphorylation of a 120 kDa protein. Cell Regul 2:951–964

Gupta S, Weiss A, Kumar G, Wang S, Nel A (1994) The T-cell antigen receptor utilizes Lck, Raf-1, and MEK-1 for activating mitogen-activated protein kinase. Evidence for the existence of a second protein kinase C-dependent pathway in an Lck-negative Jurkat cell mutant. J Biol Chem 269:17349–17357

Hatai M, Hashi H, Mogi A, Soga H, Yokota J, Yaoi Y (1994) Stimulation of tyrosine- and serine-phosphorylation of focal adhesion kinase in mouse 3T3 cells by fibronectin and fibroblast growth factor. FEBS Lett 350:113–116

Hildebrand JD, Schaller MD, Parsons JT (1993) Identification of sequences required for the efficient localization of the focal adhesion kinase, pp125FAK, to cellular focal adhesions. J Cell Biol 123:993–1005

Horak ID, Gress RE, Lucas PJ, Horak EM, Waldmann TA, Bolen JB (1991) T-lymphocyte interleukin 2-dependent tyrosine protein kinase signal transduction involves the activation of p56lck. Proc Natl Acad Sci USA 88:1996–2000

Hughes PE, Renshaw MW, Pfaff M, Forsyth J, Keivens VM, Schwartz MA, Ginsberg MH (1997) Suppression of integrin activation: a novel function of a ras/raf-initiated MAP kinase pathway. Cell 88:521–530

Hynes RO (1992) Integrins: versatility, modulation, and signaling in cell adhesion. Cell 69:11–25

Illc D, Furuta Y, Kanazawa S, Takeda N, Sobue K, Nakatsuji N, Nomura S, Fujimoto J, Okada M, Yamamoto T et al. (1995) Reduced cell motility and enhanced focal adhesion contact formation in cells from FAK-deficient mice. Nature 377:539–544

Ishino M, Ohba T, Sasaki H, Sasaki T (1995) Molecular cloning of a cDNA encoding a phosphoprotein, Efs, which contains a Src homology 3 domain and associates with Fyn. Oncogene 11:2331–2338

Johansson MW, Larsson E, Luning B, Pasquale EB, Ruoslahti E (1994) Altered localization and cytoplasmic domain-binding properties of tyrosine phosphorylated beta 1 integrin. J Cell Biol 126:1299–1309

Kornberg LJ, Earp HS, Turner CE, Prockop C, Juliano RL (1991) Signal transduction by integrins: increased protein tyrosine phosphorylation caused by clustering of beta 1 integrins. Proc Natl Acad Sci USA 88:8392–8396

Lipfert L, Haimovich B, Schaller MD, Cobb BS, Parsons JT, Brugge JS (1992) Integrin-dependent phosphorylation and activation of the protein tyrosine kinase pp125FAK in platelets. J Cell Biol 119:905–912

Maguire JE, Danahey KM, Burkly LC, van Seventer GA (1995) T cell receptor- and beta 1 integrin-mediated signals synergize to induce tyrosine phosphorylation of focal adhesion kinase (pp125FAK) in human T cells. J Exp Med 182:2079–2090

Matsuyama T, Yamada A, Kay J, Yamada KM, Akiyama SK, Schlossman SF, Morimoto C (1989) Activation of CD4 cells by fibronectin and anti-CD3 antibody. A synergistic effect mediated by the VLA-5 fibronectin receptor complex. J Exp Med 170:1133–1148

Minegishi M, Tachibana K, Sato T, Iwata S, Nojima Y, Morimoto C (1996) Structure and function of Cas-L, a 105-kD Crk-associated substrate-related protein that is involved in beta 1 integrin-mediated signaling in lymphocytes. J Exp Med 184:1365–1375

Morimoto C, Letvin NL, Boyd AW, Hagan M, Brown HM, Kornacki MM, Schlossman SF (1985) The isolation and characterization of the human helper inducer T cell subset. J Immunol 134:3762–3769

Nojima Y, Humphries MJ, Mould AP, Komoriya A, Yamada KM, Schlossman SF, Morimoto C (1990) VLA-4 mediates CD3-dependent CD4+ T cell activation via the CS1 alternatively spliced domain of fibronectin. J Exp Med 172:1185–1192

Nojima Y, Rothstein DM, Sugita K, Schlossman SF, Morimoto C (1992) Ligation of VLA-4 on T cells stimulates tyrosine phosphorylation of a 105-kD protein. J Exp Med 175:1045–1053

Nojima Y, Tachibana T, Sato T, Schlossman SF, Morimoto C (1995) Focal adhesion kinase ($pp125^{FAK}$) is tyrosine phosphorylated after engagement of a4b1 and a5b1 integrins on human T-lymphoblastic cells. Cell Immunol 161:8–13

Polte TR, Hanks SK (1995) Interaction between focal adhesion kinase and Crk-associated tyrosine kinase substrate p130Cas. Proc Natl Acad Sci USA 92:10678–10682

Polte TR, Naftilan AJ, Hanks SK (1994) Focal adhesion kinase is abundant in developing blood vessels and elevation of its phosphotyrosine content in vascular smooth muscle cells is a rapid response to angiotensin II. J Cell Biochem 55:106–119

Rozengurt E (1995) Convergent signalling in the action of integrins, neuropeptides, growth factors and oncogenes. Cancer Surv 24:81–96

Sabe H, Hata A, Okada M, Nakagawa H, Hanafusa H (1994) Analysis of the binding of the Src homology 2 domain of Csk to tyrosine phosphorylated proteins in the suppression and mitotic activation of c-Src. Proc Natl Acad Sci USA 91:3984–3988

Sakai R, Iwamatsu A, Hirano N, Ogawa S, Tanaka T, Mano H, Yazaki Y, Hirai H (1994) A novel signaling molecule, p130, forms stable complexes in vivo with v-Crk and v-Src in a tyrosine phosphorylation-dependent manner. EMBO J 13:3748–3756

Salomon DR, Mojcik CF, Chang AC, Wadsworth S, Adams DH, Coligan JE, Shevach EM (1994) Constitutive activation of integrin alpha 4 beta 1 defines a unique stage of human thymocyte development. J Exp Med 179:1573–1584

Sanchez MP, Campanero MR, Balboa MA, Sanchez MF (1993) Co-clustering of beta 1 integrins, cytoskeletal proteins, and tyrosine- phosphorylated substrates during integrin-mediated leukocyte aggregation. J Immunol 151:3817–3828

Sato T, Tachibana K, Nojima Y, D'Avirro N, Morimoto C (1995) Role of the VLA-4 molecule in T cell costimulation. Identification of the tyrosine phosphorylation pattern induced by the ligation of VLA-4. J Immunol 155:2938–2947

Schaller MD, Borgman CA, Cobb BS, Vines RR, Reynolds AB, Parsons JT (1992) pp125FAK a structurally distinctive protein-tyrosine kinase associated with focal adhesions. Proc Natl Acad Sci USA 89:5192–5196

Schaller MD, Hildebrand JD, Shannon JD, Fox JW, Vines RR, Parsons JT (1994) Autophosphorylation of the focal adhesion kinase, pp125FAK, directs SH2-dependent binding of pp60src. Mol Cell Biol 14:1680–1688

Schlaepfer DD, Hanks SK, Hunter T, van der Geer P (1994) Integrin-mediated signal transduction linked to Ras pathway by GRB2 binding to focal adhesion kinase. Nature 372:786–791

Schwartz MA (1993) Signaling by integrins: implications for tumorigenesis. Cancer Res 53:1503–1506

Schwartz MA, Lechene C, Ingber DE (1991) Insoluble fibronectin activates the Na/H antiporter by clustering and immobilizing integrin alpha 5 beta 1, independent of cell shape. Proc Natl Acad Sci USA 88:7849–7853

Seufferlein T, Rozengurt E (1994a) Lysophosphatidic acid stimulates tyrosine phosphorylation of focal adhesion kinase, paxillin, and p130. Signaling pathways and cross-talk with platelet-derived growth factor. J Biol Chem 269:9345–9351

Seufferlein T, Rozengurt E (1994b) Sphingosine induces p125FAK and paxillin tyrosine phosphorylation, actin stress fiber formation, and focal contact assembly in Swiss 3T3 cells. J Biol Chem 269:27610–27617

Shattil SJ, Brugge JS (1991) Protein tyrosine phosphorylation and the adhesive functions of platelets. Curr Opin Cell Biol 3:869–879

Shimizu Y, Shaw S (1991) Lymphocyte interactions with extracellular matrix. FASEB J 5:2292–2299

Silvennoinen O, Witthuhn BA, Quelle FW, Cleveland JL, Yi T, Ihle JN (1993) Structure of the murine Jak2 protein-tyrosine kinase and its role in interleukin 3 signal transduction. Proc Natl Acad Sci USA 90:8429–8433

Songyang Z, Shoelson SE, Chaudhuri M, Gish G, Pawson T, Haser WG, King F, Roberts T, Ratnofsky S, Lechleider RJ et al. (1993) SH2 domains recognize specific phosphopeptide sequences. Cell 72:767–778

Sultan C, Plantavid M, Bachelot C, Grondin P, Breton M, Mauco G, Levy TS, Caen JP, Chap H (1991) Involvement of platelet glycoprotein IIb-IIIa (alpha IIb-beta 3 integrin) in thrombin-induced synthesis of phosphatidylinositol 3′,4′-bisphosphate. J Biol Chem 266:23554–23557

Sun XJ, Rothenberg P, Kahn CR, Backer JM, Araki E, Wilden PA, Cahill DA, Goldstein BJ, White MF (1991) Structure of the insulin receptor substrate IRS-1 defines a unique signal transduction protein. Nature 352:73–77

Tachibana K, Sato T, D'Avirro N, Morimoto C (1995) Direct association of pp125FAK with paxillin, the focal adhesion- targeting mechanism of pp125FAK. J Exp Med 182:1089–1099

Williams DA, Rios M, Stephens C, Patel VP (1991) Fibronectin and VLA-4 in haematopoietic stem cell-microenvironment interactions. Nature 352:438–441

Wood CK, Turner CE, Jackson P, Critchley DR (1994) Characterisation of the paxillin-binding site and the C-terminal focal adhesion targeting sequence in vinculin. J Cell Sci 107:709–717

Xing Z, Chen HC, Nowlen JK, Taylor SJ, Shalloway D, Guan JL (1994) Direct interaction of v-Src with the focal adhesion kinase mediated by the Src SH2 domain. Mol Biol Cell 5:413–421

Yamada A, Nikaido T, Nojima Y, Schlossman SF, Morimoto C (1991a) Activation of human CD4 T lymphocytes Interaction of fibronectin with VLA-5 receptor on CD4 cells induces the AP-1 transcription factor. J Immunol 146:53–56

Yamada A, Nojima Y, Sugita K, Dang NH, Schlossman SF, Morimoto C (1991b) Cross-linking of VLA/CD29 molecule has a co-mitogenic effect with anti-CD3 on CD4 cell activation in serum-free culture system. Eur J Immunol 21:319–325

Zachary I, Sinnett SJ, Rozengurt E (1992) Bombesin, vasopressin, and endothelin stimulation of tyrosine phosphorylation in Swiss 3T3 cells Identification of a novel tyrosine kinase as a major substrate. J Biol Chem 267:19031–19034

Zhang C, Lambert MP, Bunch C, Barber K, Wade WS, Krafft GA, Klein WL (1994) Focal adhesion kinase expressed by nerve cell lines shows increased tyrosine phosphorylation in response to Alzheimer's A beta peptide. J Biol Chem 69:25247–25250

Zhang X, Wright CV, Hanks SK (1995) Cloning of a Xenopus laevis cDNA encoding focal adhesion kinase (FAK) and expression during early development. Gene 160:219–222

Zhang Z, Vuori K, Wang H, Reed JC, Ruoslahti E (1996) Integrin activation by R-ras. Cell 85:61–69

Functions of α4- and β7-Integrins in Hematopoiesis, Lymphocyte Trafficking and Organ Development

N. Wagner[1,2] and W. Müller[1]

1 Introduction 23
2 Structure and Expression of α4- and β7-Integrins 24
3 α4- and β1-Integrins are Crucial for Hematopoiesis 25
4 Compartmentalization of the Immune System in the Gut Is β7-Integrin-Mediated 27
5 Placental and Cardiac Development Is α4-Integrin-Dependent 29
6 Conclusion 30
References 30

1 Introduction

The family of integrins comprise more than 20 αβ heterodimeric membrane bound glycoproteins that mediate cell-cell and cell-extracellular matrix contacts. To date, 16 α and eight β subunits assemble noncovalently in a variety of combinations. In most instances, α chains associate with only one β chain while β chains are promiscuous. However, the α4 and αv subunits both associate with more than one β chain. Subfamilies of integrins are named according to the β chain used. The largest subfamily comprises the β1-integrins consisting of at least nine members. The integrins function as adhesion molecules as well as signal transducers by binding to a number of ligands (Clark and Brugge 1995; Zimmermann et al. 1996). Such ligands are either extracellular matrix proteins like fibronectin, laminin and vitronectin or members of the Ig superfamily like the intercellular adhesion molecules (ICAMs), the vascular cell adhesion molecule (VCAM-1) and the mucosal addressin cell adhesion molecule (MAdCAM-1).

The strength of integrin-mediated adhesive bonds is tightly regulated by alteration of the affinity for the respective ligands. Affinity changes are dependent upon cellular activation mechanisms that may result in conformational changes of integrins (Hughes et al. 1996; Zhang et al. 1996). A large set of functions during ontogeny and adult life has been assigned to integrins. Specifically targeted mutations of integrin genes in the mouse have been helpful in understanding the

[1]Institute for Genetics, University of Cologne, Weyertal 121, 50931 Köln, Germany
[2]Dept. of Pediatrics, University of Bonn, Bonn, Germany

physiological significance of integrins (HYNES 1996). So, integrins have crucial roles in angiogenesis and organogenesis, e.g. of placenta, skin, and kidney, hematopoiesis, and lymphocyte migration.

For further information on characteristics of integrin structure, regulation and function, a number of excellent reviews have recently been published (HYNES 1992; CARLOS and HARLAN 1994; DIAMOND and SPRINGER 1994; KILGER and HOLZMANN 1995). Herein, we review what has been learned about the role of α4- and β7-integrins in development based on the most recent published gene targeting experiments. Particularly, the function of these integrin molecules for the development of hematopoietic cells and for the compartmentalization of the immune system will be emphasized. α4- and β7-integrins may also be critical for the pathophysiology of a number of inflammatory diseases, a role that is addressed in chapters 4 and 5 of this volume of *Current Topics in Microbiology and Immunology*.

2 Structure and Expression of α4- and β7-Integrins

The α4 subunit either associates with the β1 or the β7 subunit and the β7 subunit can be expressed with the αE subunit in addition to the α4 subunit (HEMLER et al. 1990; HOLZMANN and WEISSMAN 1989; SHAW et al. 1994). The mature α4 subunit comprises 999 amino acids and has a molecular mass of 150 kDa. It can be cleaved to fragments of 80 and 70 kDa without losing its adhesive function (TEIXIDO et al. 1992). The β1 subunit and the β7 subunit have molecular masses of 116 kDa and 100 kDa, respectively. The αE subunit, so far known to form a heterodimer only with β7, has a molecular mass of 150 kDa (CERF-BENSUSSAN et al. 1992).

A number of studies have investigated the expression of the α4- and the β7-integrins in order to analyze the possible functions of these integrins. α4β1 and α4β7 surface expression is detected on lymphocytes. Specifically, most of the naive $CD4^+$ and $CD8^+$ T cells express moderate levels of both integrins, α4β1 and α4β7 (ERLE et al. 1994; PICKER et al. 1993). By contrast, following activation of lymphocytes the expression of α4-integrins increases and activated lymphocytes differentially up-regulate either α4β1 or α4β7. The latter is mainly found on lymphocytes populating the gut associated lymphoid tissue (SCHWEIGHOFFER et al. 1993; ANDREW et al. 1996). With respect to B cells the α4β1- and the α4β7-integrins are also expressed constitutively on most B cells in peripheral blood (POSTIGO et al. 1993). In lymphoid organs like spleen, peripheral lymph nodes, mesenteric lymph nodes and Peyer's patches two populations of B cells are evident which are either $β7^-$ or $β7^+$ (ERLE et al. 1994; ANDREW et al. 1996). The $β7^+$ B cells express even higher amounts of β7 than the T cells within these lymphoid organs.

Macrophages, mast cells, NK cells and eosinophils express α4β7 while monocytes, eosinophils, basophils, hematopoietic progenitor cells, and mast cells express α4β1 on their surfaces (KILGER and HOLZMANN 1995). Interestingly, neutrophils neither express α4β1 nor α4β7. αEβ7 but little α4β7 is almost exclusively found on a specific subset of mucosal T cells but not B cells, namely the intestinal intraepithelial lymphocytes (IELs) (KILSHAW and MURANT 1991; ANDREW et al. 1996); they constitute the first cellular elements of the gastrointestinal immune system encountering antigens. In addition, interdigitating dendritic cells isolated from mesenteric lymph nodes express αEβ7 (KILSHAW 1993).

α4β1 is not only found on hematopoietic cells but it is expressed on a variety of other cell lineages. So, α4β1 is expressed during embryogenesis in the heart, in smooth and skeletal muscle as well as in the neural crest (SHEPPARD et al. 1994).

3 α4- and β1-Integrins are Crucial for Hematopoiesis

The development of T and B cells takes place in the yolk sac, fetal liver, bone marrow and thymus. Within the ordered sequence of events during hematopoiesis adhesive interactions between hematopoietic progenitor cells and stromal cells take place. In addition, progenitor cells, immature lymphocytes as well as mature lymphocytes migrate between these organs to go through different developmental stages. α4- and α5-integrins are thought to mediate adhesion of hematopoietic precursor cells to stromal cells of the bone marrow based on in vitro (KERST et al. 1993) and in vivo experiments (PAPAYANNOPOULOU and NAKAMATO 1993). Furthermore, treatment of precursor cells with an anti-β1 monoclonal antibody (mAb) leads to a reduction of colony-forming units in the spleen (WILLIAMS et al. 1991). In addition, α4-integrins are differentially expressed during thymocyte maturation (SAWADA et al. 1992). These data strongly suggested that α4β1-integrin is required for lymphocyte development.

Indeed, recent gene targeting experiments in mice have provided significant insight in the in vivo role of integrins during hematopoiesis. The α4 gene inactivation in mice is lethal due to early embryonic defects in allantois-chorion fusion and cardiac defects, which will be presented more in detail below (YANG et al. 1995). Therefore, it is impossible to directly assess hematopoiesis in these mice. To circumvent this obstacle a chimera approach was utilized by which embryonic stem (ES) cells homozygous for the mutation are injected into wild-type blastocysts or into blastocysts from RAG-1 or RAG-2 deficient mice and transferred into foster mothers, thereby allowing the development of chimeric mice (CHEN et al. 1993). In chimeric mice using wild-type blastocysts the hematopoietic cells derived from the mutant ES cells compete with hematopoietic cells derived from wild-type blastocysts. By contrast, in chimeric mice using RAG (-1 or -2) deficient blastocysts the lymphocytes detected are solely derived from the mutant ES cells since the RAG mutation leads to an early developmental block of T and B cells. By the help of this approach the effect of the α4 mutation on hematopoiesis can be analysed (ARROYO et al. 1996).

α4-integrins are necessary for fetal B cell development as the number of mature B cells is greatly diminished in α4-deficient/RAG-deficient chimeric mice. In addition, B cell development seems to be compromised earlier than in the RAG mutants, namely before the pro-B cell stage. The B-1 cells in the peritoneal cavity, which comprise a distinct self renewing subset of B cells, is also compromised in α4-deficient chimeras. However, the defect in B cell development is leaky as a low number of mature B cells is found in these mice (Arroyo et al. 1996). So far functional studies of the B cells still detectable in α4-deficient chimeras are missing. The number of T cells in new-born α4-deficient/wild-type chimeras is similar to the number in control mice emphasizing that T cell development in the fetus can occur in the absence of α4-integrins. However, during the first weeks of postnatal life, the T cell numbers and particularly the number of double positive (CD4, CD8) thymocytes decreases and the thymus becomes atrophic (Arroyo et al. 1996). While prenatal T cell progenitors migrate from the yolk sac and fetal liver to the thymus, postnatally, the bone marrow replaces fetal liver with respect to lymphopoiesis. Accordingly, T cell progenitors have to migrate from the bone marrow to the thymus, which they fail to do in α4 mutants. Therefore, α4-integrins are necessary either for emigration of T cell progenitors from the bone marrow or for proper development of these cells within the bone marrow. Interestingly, intravenous injection of α4-bone marrow cells leads to reconstitution of the thymus, suggesting that in fact emigration of T cell progenitors from the bone marrow is dependent on α4-integrin function.

What about the function of α4β1- vs α4β7-integrins in this context? The data gathered from the β1 and the β7 knockouts should be informative. The β1 mutation is embryonically lethal early on due to a defect in placental implantation (Fässler and Meyer 1995). Taking into account the large number of α chains associating with β1 this effect is hard to ascribe to one particular integrin heterodimer. Again by generating chimeric mice it became obvious that β1-deficient hematopoietic progenitors fail to migrate from the yolk sac to fetal liver which prevents further hematopoiesis (Hirsch et al. 1996). Most interestingly, the progenitors derived from the yolk sac were capable of differentiating in vitro into B cells. The gene inactivation of the β7 gene led to viable offspring. In these mice T cell as well as B cell development is unaltered suggesting that β7 does not play a crucial role for lymphocyte development (Wagner et al. 1996). Finally, the αE-deficient mice also bear normal numbers of peripheral B and T cells (Parker et al., unpublished personal communication). Based on these data a number of conclusions can be drawn for the role of α4- and β7-integrins in lymphocyte development:

1. β1-integrins but not α4β1 are essential for migration of hematopoietic progenitors from the yolk sac to the fetal liver. α4β1 is essential for B cell development while other β1-integrins are not.
2. α4β1 and α4β7 may complement each other in B cell development. This would explain the observed phenotypes of the gene targeted mice, namely, that B cell development is normal in β1-deficient chimeric mice and in β7-deficient mice but it is severely impaired in α4-deficient chimeric mice.

3. Prenatal T cell development is neither dependent on α4β1 nor on α4β7 while postnatal T cell development is α4β1-dependent. In particular, the emigration of T cell progenitors from the bone marrow to the thymus may be α4β1-mediated.

The β7- and the α4-integrins are not critical for the development of hematopoietic cells other than lymphocytes like erythrocytes, myeloid cells and NK cells (Hirsch et al. 1996; Arroyo et al. 1996; Wagner et al. 1996).

4 Compartmentalization of the Immune System in the Gut Is β7-Integrin-Mediated

Numerous nutritive antigens and pathogens pass through the gut allowing a vast number of antigens to come into contact with the mucosa (Brandtzaeg 1989). The challenge for the immune system of the gut is to fight harmful pathogens while at the same time to be tolerant towards harmless nutritive antigens passing the mucosa. If to every foreign antigen in the gut an immune response would be elicited nutrition of the organism would be impossible. Therefore, the immune system has evolved tolerance mechanisms which in the gut may depend on the specific microenvironment in the gut associated lymphoid tissue (GALT) (Weiner et al. 1994).

The GALT is one compartment of the immune system. It comprises specific lymphoid organs, i.e., the Peyer's patches (PP), the lamina propria lymphocytes (LPLs) and a distinct subset of T cells, the intestinal intraepithelial lymphocytes (IELs) which drastically differ from peripheral T cells with respect to their phenotype (Brandtzaeg 1989; Sim 1995; Klein 1996). Lymphocytes that have been primed in the gastrointestinal tract tend to recirculate within this compartment to generate or to prevent an efficient immune response. Continuous trafficking of lymphocytes ensures that the highly diverse antigen receptors on the surface of lymphocytes come into physical contact with the antigens at any site of the body (Butcher and Picker 1996). Recirculation of lymphocytes supports the regionalization of the immune system which means that the immune response to an antigen may be dependent on the regional microenvironment where the contact between lymphocyte and antigen takes place. Adhesion molecules like integrins, selectins and members of the Ig superfamily mediate the trafficking of lymphocytes.

In recent years a multistep model of leukocyte trafficking has been developed: Leukocytes first randomly come into contact with the vessel wall; they start rolling along the endothelium; in the next step leukocytes are activated by chemokines, they stop and firmly adhere and then they finally transmigrate through the endothelial cells (Butcher 1991; Springer 1994). Organ specific trafficking is regulated by expression levels of adhesion molecules participating at different stages of the adhesion cascade, by the activation state of adhesion molecules as well as by the interaction of a variety of chemokines with their receptors (Carlos and Harlan 1994; Butcher and Picker 1996).

A lymphocyte adhesion molecule specific for PP which, as already mentioned, are part of the GALT has originally been identified by a mAb that inhibits binding of murine lymphocytes to high endothelial venules (HEVs) of PP (HOLZMANN and WEISSMAN 1989). The HEVs are vessels with unusual, high walled endothelial cells that are only detected in lymphoid organs and specifically function as the entry site for lymphocytes (GIRARD and SPRINGER 1995). Using the mAb binding to the leukocyte PP adhesion molecule (LPAM-1) the nature of the glycoprotein as an integrin heterodimer of the α4 and the β7 chain has been defined (HOLZMANN and WEISSMAN 1989). One year earlier the adhesion molecule on mucosal endothelial cells that is now known to be the ligand of α4β7 and is termed MAdCAM-1 was described (STREETER et al. 1988). However, other adhesion molecules like LFA-1, CD44 and L-selectin have also been ascribed a role in mediating lymphocyte migration to the gut (JALKANEN et al. 1987; HULEATT and LEFRANCOIS 1996; HAMANN et al. 1994; BARGATZE et al. 1995).

The contribution of α4- and β7-integrins to lymphocyte migration can be assessed by the results from gene targeting experiments of the respective integrin subunits as well as of the αE subunit. The β7-deficient mice are viable and do not exhibit any defect in lymphocyte development. However, the compartmentalization of the immune system in these mice is affected. The GALT is severely compromised as the PP are drastically reduced in size and cellularity and the numbers of LPLs and IELs are also diminished (WAGNER et al. 1996). The likely cause for this impaired seeding of the GALT with lymphocytes is that directed migration of β7-deficient lymphocytes to the GALT is almost abolished, as shown by short-term migration assays and the lymphocyte-endothelial cell adherence assay (Stamper Woodruff assay). In addition, transfer of $\beta7^+$ bone marrow into β7-deficient mice rescues the formation of the GALT. How do α4β7 vs αEβ7 and α4β1 function in mediating lymphocyte migration to the GALT? αE deficient mice have a reduced number of IELs and of LPLs, however the PP are normal with respect to size and cellularity (PARKER et al., unpublished personal communication). The α4-deficient chimeric mice have normal numbers of IEL but PP are drastically affected (ARROYO et al. 1996). These data lead to the following conclusions for the function of β7-integrins:

1. α4β7 is critical for lymphocyte migration to PP while αEβ7 is not involved in this process and α4β1 contributes only to a minor extent if at all.
2. α4β7 and αEβ7 are important for lymphocyte migration to the lamina propria of the intestine; the significance of α4β1 cannot be assessed since it was not examined in the α4-deficient mice.
3. αEβ7 is critical for the localization of lymphocytes to the intraepithelial site of the intestine while α4β7 and α4β1 will probably not contribute to this process. However, the localization of only a subset of IELs is mediated by αEβ7.

To define the step in the adhesion cascade that is affected by the β7 deficiency, in vivo epifluorescence video microscopy in PP HEVs was performed. Fluorescently labeled β7-deficient lymphocytes were still capable of rolling along the endothelium; however they no longer firmly adhered to the endothelial cells or transmigrated through the endothelium (WAGNER et al. 1996). Therefore, the transition from rolling of lymphocytes in PP HEVs to firm adhesion of lymphocytes to endothelial cells is mediated by β7-integrins. A thorough analysis of the velocities with which β7-deficient lymphocytes roll along the endothelium of PP HEVs demonstrated a higher speed than wild-type controls (Ley and Wagner, unpublished observations). This suggests a contribution of β7-integrins to mediating lymphocyte rolling in PP HEVs, as proposed by Bargatze and co-workers (BARGATZE et al. 1995).

Taken together α4- and β7-integrins participate in a complex regulation of organ specific lymphocyte trafficking, emphasizing the significance of different compartments of the immune system. Future studies with the herein discussed gene targeted mice will address the pathophysiological consequences of disturbed lymphocyte migration with respect to immune responses to pathogens and to inflammatory bowel disease. Furthermore, signal transduction events mediated by integrins may be important for the observed phenotypes.

5 Placental and Cardiac Development Is α4-Integrin-Dependent

The early embryonic lethality of the α4 mutation is not caused by the effects on hematopoiesis or lymphocyte trafficking but on placental and cardiac development (YANG et al. 1995). The phenotype detected in α4-deficient embryos closely resembles the phenotype observed in VCAM-1 deficient embryos (KWEE et al. 1995; GURTNER et al. 1995). VCAM-1 is the membrane bound ligand of α4β1 and their specific interaction is essential for distinct features of placental and cardiac development. α4-deficient embryos die at around embryonic day 11 due to the failure of the allantois to fuse with the chorion. The α4-integrin is expressed in the chorion and VCAM-1 is expressed in the allantois, both at the site of fusion. In some α4-deficient embryos the fusion of allantois and chorion is successful. However, those embryos die at embryonic day 11.5 due to severe hemorrhage in the heart region. Here, the α4 mutation causes an impaired formation of the epicardium and a lack of coronary vessels (YANG et al. 1995). Again, this observation is very similar to the cardiac defects detected in VCAM-1 deficient embryos (KWEE et al. 1995; GURTNER et al. 1995). Along with these findings goes the expression of α4-integrin and VCAM-1 in the heart.

That expression data alone do not suffice to conclude towards the function of a particular gene becomes evident when considering muscle development and the role of α4-integrins. α4-integrins are expressed on primary myotubes during muscle development but in α4-deficient chimeric mice no defect in skeletal muscle was observed thereby challenging the hypothesis of α4-integrin involvement in muscle development (YANG et al. 1995).

6 Conclusion

α4- and β7-integrins (α4β1, α4β7 and αEβ7) are critically involved in hematopoiesis, and/or in compartmentalization of the immune system, and/or in placental and cardiac development. The gene targeting of single integrin subunits in mice contributed significantly to our understanding of the developmental significance of the different integrin heterodimers. Future studies will utilize these mice for in vivo studies of immune function. In case of early embryonic lethality tissue specific or conditional gene targeting will help to address this obstacle (RAJEWSKY et al. 1996).

Acknowledgements. The authors thank Dr. K. Rajewsky for helpful discussions and continuous support, Drs. A. Tarakhovsky and C. Leuker for critical reading of the manuscript and Dr. C. Parker for communicating data prior to publication. N.W. has been a recipient of an AIDS grant from the Bundesministerium für Bildung und Forschung. In addition, this work was supported by a grant (Genzentrum Köln) from the Bundesministerium für Bildung und Forschung to Klaus Rajewsky and W.M. and a grant from the Deutsche Forschungsgemeinschaft (WA1127/1–1).

References

Andrew DP, Rott LS, Kilshaw PJ, Butcher EC (1996) Distribution of α4β7 and αEβ7 integrins on thymocytes, intestinal epithelial lymphocytes and peripheral lymphocytes. Eur J Immunol 26:897–905

Arroyo AG, Yang JT, Rayburn H, Hynes RO (1996) Differential requirements for α4 integrins during fetal and adult hematopoiesis. Cell 85:997–1008

Bargatze RF, Jutila MA, Butcher EC (1995) Distinct roles of L-selectin and integrins α4β7 and LFA-1 in lymphocyte homing to Peyer's patch-HEV in situ: the multistep model confirmed and refined. Immunity 3:99–108

Brandtzaeg P (1989) Overview of the mucosal immune system. In: Mestecky U, McGhee JR (eds) New strategies for oral immunization. Springer, Berlin Heidelberg New York, pp 13–25 (Current topics in microbiology and immunology, vol 146)

Butcher EC (1991) Leukocyte-endothelial cell recognition: three (or more) steps to specificity and diversity. Cell 67:1033–1036

Butcher EC, Picker LJ (1996) Lymphocyte homing and homeostasis. Science 272:60–66

Carlos TM, Harlan JM (1994) Leukocyte-endothelial adhesion molecules. Blood 84:2068–2101

Cerf-Bensussan N, Begue B, Gagnon J, Meo T (1992) The human intraepithelial lymphocyte marker HML-1 is an integrin consisting of a β7 subunit associated with a distinctive α chain. Eur J Immunol 22:273–277

Chen J, Lansford R, Stewart V, Young F, Alt FW (1993) RAG-2-deficient blastocyst complementation: an assay of gene function in lymphocte development. Proc Natl Acad Sci USA 90:4528–4532

Clark EA, Brugge JS (1995) Integrins and signal transduction pathways: the road taken. Science 268:233–239

Diamond MS, Springer TA (1994) The dynamic regulation of integrin adhesiveness. Curr Biol 4:506–517

Erle DJ, Briskin MJ, Butcher EC, Garcia Pardo A, Lazarovits AI, Tidswell M (1994) Expression and function of the MAdCAM-1 receptor, integrin α4β7 on human leukocytes. J Immunol 153:717–729

Fässler R, Meyer M (1995) Consequences of lack of β1 integrin gene expression in mice. Genes Dev 9:1896–1908

Girard JP, Springer TA (1995) High endothelial venules (HEVs): specialized endothelium for lymphocyte migration. Immunol Today 16:449–457

Gurtner GC, Davis V, Li H, McCoy MJ, Sharp A, Cybulsky MI (1995) Targeted disruption of the murine VCAM-1 gene: essential role of VCAM-1 in chorioallantoic fusion and placentation. Genes Dev 9:1–14

Hamann A, Andrew DP, Jablonski-Westrich D, Holzmann B, Butcher EC (1994) Role of α4-integrins in lymphocyte homing to mucosal tissues in vivo. J Immunol 152:3282–3292

Hemler ME, Elices MJ, Parker C, Takada Y (1990) Structure of the integrin VLA-4 and its cell-cell and cell-matrix adhesion functions. Immunol Rev 114:45–65

Hirsch E, Iglesias A, Ptocnik AJ, Hartmann U, Fässler R (1996) Impaired migration but not differentiation of haematopoetic stem cells in the absence of β1 integrins. Nature 380:171–175

Holzmann B, Weissman IL (1989) Peyer's patch-specific lymphocyte homing receptors consist of a VLA-4 like α chain associated with either of two integrin β chains, one of which is novel. EMBO J 8:1735–1741

Hughes PE, Diaz-Gonzales F, Leong L, Wu C, Mc Donald JA, Shattil SJ, Ginsberg MH (1996) Breaking the integrin hinge. A defined structural constraint regulates integrin signaling. J Biol Chem 271:6571–6574

Huleatt JW, Lefrancois L (1996) β2 integrins and ICAM-1 are involved in establishment of the intestinal mucosal T cell compartment. Immunity 5:263–273

Hynes RO (1992) Integrins: versatility, modulation, and signaling in cell adhesion. Cell 69:11–25

Hynes RO (1996) Targeted mutations in cell adhesion genes: what have we learned from them? Dev Biol 180:402–412

Jalkanen S, Bargatze RF, de los Toyos J, Butcher EC (1987) Lymphocyte recognition of high endothelium: antibodies to distinct epitopes of an 85–95KD glycoprotein antigen differentially inhibit lymphocyte binding to lymph node, mucosal, or synovial endothelial cells. J Cell Biol 105:983–990

Kerst JM, Sanders JB, Slaper-Cortenach ICM, Doorakkers MCh, Hooibrink B, van Oers RHJ, Von dem Borne AE, van der Schoot CE (1993) α4β1 and α5β1 are differentially expressed during myelopoesis and mediate the adherence of human $CD34^+$ cells to fibronectin in an activation-dependent way. Blood 81:344–351

Kilger G, Holzmann B (1995) Molecular analysis of the physiological and pathophysiological role of α4-integrins. J Mol Med 73:347–354

Kilshaw PJ (1993) Expression of the mucosal T cell integrin $\alpha m_{290}\beta 7$ by a major subpopulation of dendritic cells in mice. Eur J Immunol 23:3365–3368

Kilshaw PJ, Murant SJ (1991) Expression and regulation of β7(βp) integrins on mouse lymphocytes: relevance to the mucosal immune system. Eur J Immunol 21:2591–2597

Klein JR (1996) Whence the intestinal intraepithelial lymphocytes. J Ep Med 184:1203–1206

Kwee L, Baldwin HS, Shen HM, Stewart CL, Buck C, Buck CA, Labow MA (1995) Defective development of the embryonic and extraembryonic circulatory systems in vascular adhesion molecule (VCAM-1) deficient mice. Development 121:489–503

Papayannopoulou T, Nakamato B (1993) Peripherilization of hemopoetic progenitors in primates treated with anti-VLA-4 integrin. Proc Natl Acad Sci USA 90:9374–9378

Picker LJ, Treer JR, Nguyen M, Terstappen LW, Hogg N, Yednock T (1993) Coordinate expression of β1 and β2 integrin "activation" epitopes during T cell responses in secondary lymphoid tissue. Eur J Immunol 23:2751–2757

Postigo AA, Sanchez Mateos P, Lazarovits AI, Sanchez Madrid F, De Landazuri MO (1993) α4β7 integrin mediates B cell binding to fibronectin and vascular adhesion molecule-1: expression and function of α4 integrins on human B lymphocytes. J Immunol 151:2471–2483

Rajewsky K, Gu H, Kühn R, Betz UAK, Müller W, Schwenk F (1996) Conditional gene targeting. J Clin Invest 98:600–603

Sawada M, Nagamine J, Takeda K, Utsumi K, Kosugi A, Tatsumi Y, Hamaoka T, Miyake K, Nakajima K, Watanabe T, Sakakibara S, Fujiwara H (1992) Expression of VLA-4 on thymocytes: maturation stage-associated transition and its correlation with their capacity to adhere to thymic stromal cells. J Immunol 149:3517–3524

Schweighoffer T, Tanaka Y, Tidswell M, Erle DJ, Horgan KJ, Ginther Luce GE, Lazarovits AI, Buck D, Shaw S (1993) Selective expression of integrin α4β7 on a subset of human CD4+ memory T cells with hallmark of gut-tropism. J Immunol 151:717–729

Shaw SK, Cepek KL, Murphy EA, Russell GJ, Brenner MK, Parker CM (1994) Molecular cloning of the human mucosal lymphocyte integrin αE subunit. J Biol Chem 269:6061–6025

Sheppard AM, Onken MD, Rosen GD, Noakes PG, Dean DC (1994) Expanding roles for α4 integrins and its ligands in development. Cell Adhes Commun 2:27–43

Sim GK (1995) Intraepithelial lymphocytes and the immune system. Adv Immunol 58:297–343

Springer T (1994) Traffic signals for lymphocyte recirculation and leukocyte emigration: the multistep paradigm. Cell 76:301–314

Streeter PR, Berg EL, Rouse BT, Bargatze RF, Butcher EC (1988) A tissue-specific endothelial cell molecule involved in lymphocyte homing. Nature 331:41–46

Teixido J, Parker CM, Kassner PD, Hemler ME (1992) Functional and structural analysis of VLA-4 integrin α4 subunit cleavage. J Biol Chem 267:1786–1791

Wagner N, Löhler J, Kunkel EJ, Ley K, Leung E, Krissansen G, Rajewsky K, Müller W (1996) Critical role for β7 integrins in formation of the gut associated lymphoid tissue. Nature 382:366–370

Weiner HL, Friedman A, Miller A, Khoury SJ, Al-Sabbagh A, Santos L, Sayegh M, Nussenblatt RB, Trentham DE, Hafler DA (1994) Oral tolerance: immunologic mechanisms and treatment of animal and human organ-specific autoimmune diseases by oral administration of autoantigens. Annu Rev Immunol 12:809–837

Williams DA, Rios M, Stephens C, Patel VP (1991) Fibronectin and VLA-4 in haematopoetic stem cell-microenvironment interactions. Nature 352:438–441

Yang JT, Rayburn H, Hynes RO (1995) Cell adhesion events mediated by α4 integrins are essential in placental and cardiac development. Development 121:549–560

Zhang Z, Vuori K, Wang GG, Reed JC, Ruoslahti E (1996) Integrin activation by R-ras. Cell 85:61–69

Zimmermann GA, McIntyre TM, Prescott SM (1996) Adhesion and signaling in vascular cell-cell interactions. J Clin Invest 98:1699–1702

Regulation of Integrin Function by Inside-Out Signaling Mechanisms

W. KOLANUS and L. ZEITLMANN

1 Introduction 33

2 Activation Dependence of Integrin Adhesiveness 34

3 The Role of Integrin Cytoplasmic Domains in Inside-Out Signal Transduction 35

4 Involvement of Cytoskeletal Interactions in the Activation Pathway of Integrins 37

5 Signal Transduction Pathways Which Control Integrin Adhesiveness 38

6 Proximal Regulatory Molecules at the Interface of Integrin Cytoplasmic Domains and Upstream Signaling Pathways 40

7 Modulation of Integrin Adhesiveness – Potential Mechanisms 42

References 46

1 Introduction

Coordinated cell-cell and cell-substratum interactions are crucial for the differentiation, organization and diverse somatic functions of multicellular organisms. Several families of molecules have evolved to serve such specialized purposes such as cell migration in development and in the immune response. One of these families, the integrins, plays a prominent role in the attachment and detachment of cells to their surrounding matrix, the adhesion of platelets to fibrinogen, the coupling of lymphocytes to antigen presenting cells and the phagocytosis of complement opsonized targets by myelomonocytic phagocytes in the immune system. At least 21 different integrin receptors have been characterized to date, all of which are heterodimeric transmembrane proteins that comprise an α subunit which is non-covalently associated with a β subunit at the cell surface. The family was originally characterized by structurally related β subunits that formed a series of heterodimers with distinct α subunits. However, recent findings indicate that there are several important exceptions and that more elaborate schemes are required to describe the systematic structural organization of these receptors. A number of comprehensive reviews on this issue have appeared recently (DIAMOND and SPRINGER 1994; HEMLER 1990; HYNES 1992; SPRINGER 1990). Integrins bind to a diverse array of

Laboratorium für Molekulare Biologie, Genzentrum der Universität München, Feodor-Lynen Straße 25, 81377 Munich, Germany ligands including extracellular matrix (ECM) proteins, plasma proteins which mediate hemostasis and complement activation (fibrinogen and ic3b), and integral membrane proteins. Many integrins bind to more than one ligand and vice versa, using either identical or distinct binding sites.

The biological importance of integrins is underscored by the expression of multiple integrins on almost every cell type in the body, by the embryonic lethal phenotype of mice lacking expression of the integrin β1 subunit and associated α subunits (Yang et al. 1993, 1995), by the severe immunodeficiency observed in patients lacking the β2 subunit (Springer 1990), and by the defective immune functions of mice that lack β1-integrin expression in hematopoietic cells (Hirsch et al. 1996).

It appears obvious that the interactions between integrins and their ligands have to be highly regulated. Spatially and temporally regulated expression of distinct receptor-ligand pairs (e.g., activated expression of integrin ligands on endothelial cells in the course of inflammation) is one strategy to achieve such functional specificity. However, cell surface expression of an integrin alone usually is not sufficient for adhesion. It is now described for many cell types that integrin mediated cell binding to extracellular ligands can be enhanced dramatically by stimuli, such as PMA, which activate intracellular signal transduction cascades, although the density of cell surface expression appears unaltered or almost unaltered at the same time. Integrins must thus be "activated" from the cytoplasm in order to bind to their ligands, and this phenomenon has been termed "inside-out signaling." This regulation is not only important for cell adhesion, it is also actively involved in the assembly of the extracellular matrix (Wu et al. 1995). We have just begun to understand how signals that emanate in the course of cellular activation events regulate the functional properties of these important adhesion receptors. It is currently accepted that the adhesion of integrins to their ligands may be regulated by changing the affinity state of the integrin directly, most likely through induced conformational changes, or by alterations in avidity which may be mediated by, e.g., receptor aggregation. Throughout this review we will refer to these phenomena as integrin "activation," because the actual structural basis of these mechanisms is not precisely known. Several excellent reviews have appeared in the course of the last couple of years which cover this topic broadly (Dedhar and Hannigan 1996; Diamond and Springer 1994; Ginsberg et al. 1992; Hynes 1992; Lub et al. 1995). This overview attempts to highlight some of the recent exciting findings which may help to develop mechanistic concepts of these processes.

2 Activation Dependence of Integrin Adhesiveness

The affinity or avidity of integrins for their extracellular ligands can change in response to cytoplasmic signals initiated by the stimulation of a wide variety of cellular receptors. All integrin families have been shown to undergo activation (i.e., become susceptible for ligand interactions). Here we will focus mainly on β1, β2 and β3-integrins.

Activation of leukocytes with stimulating agents like phorbol esters (DUSTIN and SPRINGER 1989), isolated lipids from activated cells of hematopoietic origin (HERMANOWSKI et al. 1992; LEE et al. 1994), calcium ionophores (LUB et al. 1995; SHIMIZU and HUNT 1996), chemoattractants, or by the aggregation of functionally relevant surface receptors – such as the antigen receptor/CD3 complex (DUSTIN and SPRINGER 1989), or CD2 (VAN KOOYK et al. 1989) – causes αLβ2 (LFA-1, CD11a/18) to bind to purified ICAM-1 within minutes. Dependent on the cell type and stimulus the increase in adhesion may be transient or permanent. These regulatory pathways probably play important roles in the conjugation of lymyphocytes with antigen presenting cells as well as in leukocyte-endothelial cell interactions. A similar activation requirement has been observed for the αMβ2 receptor that is found on the surface of myeloid cells. However, an auxiliary regulatory mechanism is present in these cells: an up-regulation of αMβ2 at the cell surface occurs after stimulation which stems from the translocation of secretory granules that contain intracellular pools of αMβ2-integrin (DIAMOND and SPRINGER 1994).

On resting platelets αIIbβ3-integrin does not bind to fibrinogen. Activation occurs after vascular injury and the release of inflammatory mediators which result in platelet aggregation. Binding of αIIbβ3 to fibrinogen in vitro can be stimulated by agonists such as thrombin or platelet activating factor (PAF) (GINSBERG et al. 1992). Ectopic expression of the platelet integrin in B cells resulted in fibrinogen binding which was up-regulated by phorbol ester, suggesting that proximal elements of integrin activation pathways may be shared between platelets and B cells (LOH et al. 1995).

The β1-integrins on hematopoietic cells bind to endothelial cells and to ECM proteins which makes them important regulators of diapedesis and cell migration. Phorbol esters, as well as the stimulation of the CD2, CD3, CD7, CD28 and CD31 surface receptors on lymphocytes, chemotactic proteins which bind to seven transmembrane receptors, or the activation of the platelet derived growth factor (PDGF) receptor on nonhematopoietic cells induce β1-integrin binding to fibronectin, laminin, or VCAM-1 (DIAMOND and SPRINGER 1994; SHIMIZU and HUNT 1996).

3 The Role of Integrin Cytoplasmic Domains in Inside-Out Signal Transduction

The cytoplasmic domains of integrins became the focus of research efforts which aimed at understanding the mechanism of integrin activation from inside the cell. Compelling evidence suggested that the intracellular portions of both α (CHAN et al. 1992; FILARDO and CHERESH 1994; KASSNER and HEMLER 1993; KASSNER et al. 1994; KAWAGUCHI and HEMLER 1993; O'TOOLE et al. 1991) and β chains (CHEN et al. 1992; HAYASHI et al. 1990; HIBBS et al. 1991b; PASQUALINI and HEMLER 1994) participate in this process. Phosphorylation of the cytoplasmic domains occurs and has been proposed to contribute to activation events (VALMU and GAHMBERG 1995). However, other studies argue against a direct role of receptor phosphorylation in the promotion of cell adhesion (HIBBS et al. 1991a; O'TOOLE et al. 1995).

Mutational analysis of the β2 cytoplasmic domain in cos cells revealed that a series of three consecutive threonines (TTT 758–760) was important for basal binding of αLβ2 to ICAM-1. However, this mutant (TTT/AAA) still responded to activation mediated by phorbol esters (HIBBS et al. 1991a). Chimeric receptors which consisted of the αLβ2 cytoplasmic domains fused to the transmembrane and extracellular domains of the platelet integrin αIIbβ3 were used to study the role of the TTT 758–760 region in CHO cells. This study confirmed the importance of this element in cell adhesion and suggested that it links the β2-integrins to the cytoskeleton. However, a high affinity state of the chimera was induced and locked by deletion of the highly conserved membrane proximal GFFKR region of the α chain cytoplasmic domain, even in the presence of the TTT/AAA mutation of the β chain. Thus, high affinity ligand binding and cell adhesion appeared to be dissociable (PETER and O'TOOLE 1995). The GFFKR deletion mutant has been described to result in the activation of β3- or β1-integrins as well, which underscores the general importance of this α chain element in the regulation of integrin-ligand interactions (O'TOOLE et al. 1994).

A number of studies proved the importance of the β1 or β3 cytoplasmic domains in the regulation of inside-out signaling. These studies made use of a monoclonal antibody (PAC-1) which recognizes an inducible extracellular epitope of the αIIbβ3-integrin thought to be indicative of a high affinity binding potential of αIIbβ3 for fibrinogen. Isolated β1 or β3 cytoplasmic domains fused to heterologous transmembrane and extracellular elements were overexpressed in cos- and CHO cells and found to interfere with the activation of a co-overexpressed integrin chimera, suggesting that the isolated cytoplasmic domains are sequestering intracellular elements which are necessary for activation (CHEN et al. 1994a). Two α subunit cytoplasmic domains (α5, αIIb) were inactive in this system and, furthermore, a point mutant of the β3 cytoplasmic domain (S752P) which had previously been shown to block the expression of the high affinity epitope of αIIbβ3 consequently lacked this inhibitory potential (CHEN et al. 1994b). Chimeric integrin receptors were also used to investigate whether pairing the β1 cytoplasmic domain with different α chain intracellular elements would affect the activation potential. Interestingly, certain combinations (β1 with α2, α6A or α6B) were permissive, while others (β1with αM, αL or αV) were inhibitory, again reinforcing the idea that both α and β chain cytoplasmic domains are involved in the control of integrin activation (O'TOOLE et al. 1994).

The β3 cytoplasmic domain was further mapped for regulatory elements. It was found that the deletion of a region comprising eight membrane proximal amino acids resulted in activation, an effect which was reminiscent of the GFFKR deletion of the α chain (HUGHES et al. 1995). With the help of an elegant mutational approach both α and β chain membrane proximal sequences were subsequently shown to be involved in integrin activation. These sequences were proposed to form a charge interaction dependent "hinge" which may define the default inactive state (HUGHES et al. 1996).

The β chain cytoplasmic domains of integrins contain a conserved NPXF/Y motif that has been known for some time to connect cell surface receptors to a classical internalization pathway. This motif has also been implicated to be involved in the activation of integrin chimeras which contain wild-type or mutated β1 or β3 cytoplasmic domains. Accordingly, differential expression of this motif on β1 splice variants appeared to correlate with their intrinsic activation potential. Phosphorylation of the motif did not appear to account for the observed effects (O'Toole et al. 1995).

The contribution of sequences within the β7 cytoplasmic domain to affinity modulation have been mapped. The data presented in this study resemble the findings for other integrins in that COOH-terminal sequences are required for inside-out regulation whereas interference with NH_2-terminal residues results in constitutive activation (Crowe et al. 1994).

Cytoplasmic elements which have been shown to be important for the modulation of integrin function are summarized in Fig. 1.

4 Involvement of Cytoskeletal Interactions in the Activation Pathway of Integrins

The integrin cytoplasmic domains undergo numerous interactions with cytoskeletal proteins, some of which may contribute to activation of cell adhesion.

2.

β3 K**LLItiHDR**k EFAKFAeEra rakWdtar**NP LY**KeAt**s**Tft NitYrgt

1. 3.

β1 K**LLmiIHDR**R E**FAKFEkEkm** nAkWdtge**NP iY**KsAvtTvv NpkYegk

4. 5.

β2 KaLIhlsDl**R EYrrFEkEk**l ksqWnnd.NP LFKsA**ttT**vm NpkFaes

6.

e.g. αL KV**GFFKR**NLKEKMEAGRGV PNGIPAEDSEQLASGQEAGD PGCLKPLHEKDSESGGGKD

Fig. 1. Elements in the cytoplasmic domains which may be important for the regulation of integrin adhesion, modified from (Hughes et al. 1995). Conserved residues within the β chains are in *uppercase letters*. *1*, *6*, Membrane proximal elements of α and β chains the deletion of which leads to a constitutively active phenotype. The GFFKR region of some α chains contains the binding site for calreticulin. *4*, α-actinin binding sites of β1 and β2 integrins. *3*, The NPXY region, implicated in activation of β1 and β3 integrins, which is also important for internalization of surface receptors. The binding site for talin overlaps with this motif. *2*, *5*, Distal residues, necessary for the adhesion of β3 and β2 integrins. References are given in the text

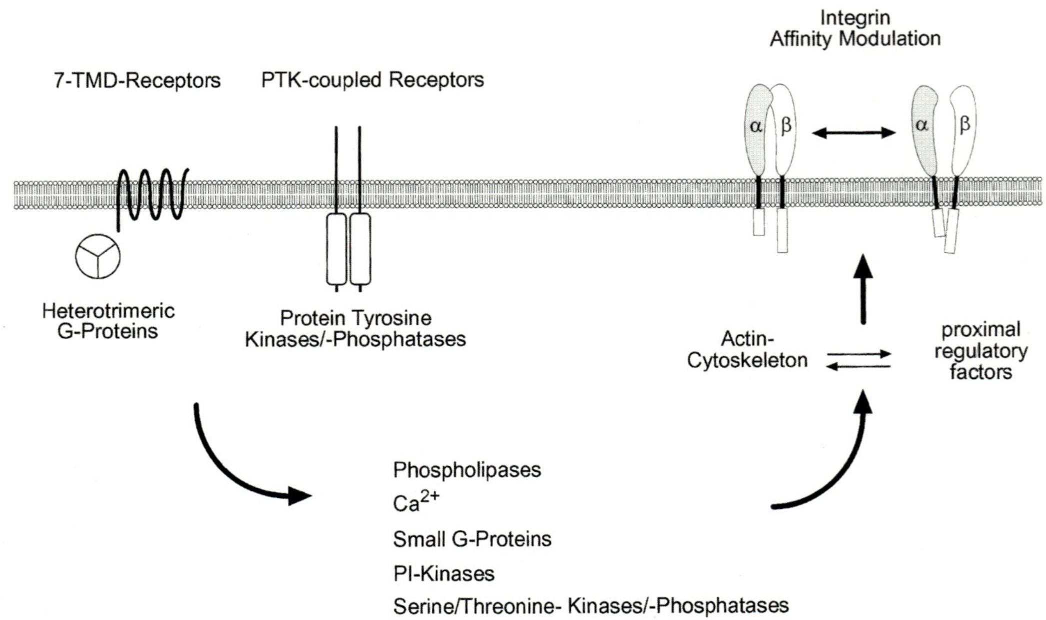

Fig. 2. Signaling pathways which appear to contribute to the activation of integrin adhesiveness

Actin binding proteins like α-actinin and talin bind directly to the β1A cytoplasmic domain (HORWITZ et al. 1986; LEWIS and SCHWARTZ 1995; OTEY et al. 1990, 1993). α-Actinin also binds to the β2 cytoplasmic tail in vitro and the proteins can be coprecipitated from detergent lysates of neutrophilic granulocytes. Notably, association of α-actinin with the β2-integrins appeared to be enhanced in fMet-Leu-Pro (fMLP) activated neutrophils (PAVALKO and LAROCHE 1993). The interaction sites within the cytoplasmic domains have been mapped by in vitro and in vivo studies and show partial correlation with each other or with activation events (DEDHAR and HANNIGAN 1996). α-Actinin and filamin bind overlapping but distinct motifs that comprise highly charged conserved elements present within all integrin cytoplasmic domains (SHARMA et al. 1995). Talin and also α-actinin can apparently make contact with elements that contain conserved NPXY/F motifs which have been found to be involved in the activation of adhesion (OTEY et al. 1993) (also see previous paragraph).

5 Signal Transduction Pathways Which Control Integrin Adhesiveness

Many groups have tried to identify and characterize signaling pathways upstream of the cytoplasmic domains of integrins which may contribute to the regulation of cell adhesion. An outline of our current knowledge is shown in Fig. 2. Methodological approaches include the use of specific inhibitors of, e.g, protein kinases and phosphatases, as well as the overexpression of dominant negative or constitutively active versions of proteins capable of transducing signals.

Cell surface receptors that couple to nonreceptor tyrosine kinases are known to activate integrin dependent adhesion pathways in leukocytes, whereas seven transmembrane proteins, like thrombin- or chemokine receptors, couple to downstream elements through the activation of heterotrimeric G proteins. The common denominator is that both types of receptors stimulate pathways involving activation of phospholipases of the B and C classes and subsequent phosphatidylinositol lipid breakdown, the results of which are a rise in intracellular Ca^{2+} concentration, a necessary requirement for induced changes of cell adhesion, and activation of protein kinase C (PKC) by diacylglycerol (HYNES 1992; LUB et al. 1995).

Isoforms of PKC were early targets of investigation because the observation that phorbol esters were such potent activators of cell adhesion prompted the hypothesis that PKC might be directly involved in these processes. PKC inhibitors such as staurosporine, calphostin C or GF109203X apparently had opposing effects on integrin mediated adhesion of different cell types, although they mostly appeared to abrogate integrin activation (DIAMOND and SPRINGER 1994; HAUSS et al. 1993; KLEMKE et al. 1994; PACIFICI et al. 1994; VAN LEEUWEN et al. 1994). This may partially be due to the fact that these cell types express alternative PKC isoforms which may play differential roles in integrin activation. It was proposed that PKC isoforms are differentially translocated to the plasma membrane in HeLa cells adhering to collagen and in leukocytes (Chun et al. 1996; Kiley and Parker 1995). Furthermore, other serine/threonine kinases such as calmodulin kinase may be involved in the regulation of integrin adhesion (Eguchi and Horikoshi 1996). PKCs may phosphorylate β chain cytoplasmic domains directly in vivo, but a correlation of this modification with integrin affinity modulation has also been discounted (Hibbs et al. 1991a; Valmu and Gahmberg 1995; van Willigen et al. 1996). Nonetheless, a number of studies argue for a plausible involvement of PKC as upstream regulator of β1, β2 and β3-integrin activation. The direct target(s) of PKC in these pathways remain mostly obscure. Recently, a genetic approach was reported which aimed at identifying factors for the regulation of leukocyte adhesion downstream of PKC. A mutant cell line was isolated which had defects in cell adhesion but an intact IL-2 promoter activation pathway (Mobley et al. 1996). Interestingly, the phosphorylation of pleckstrin, the major PKC substrate in platelets, is correlated with αIIbβ3 activation (Gabbeta et al. 1996). When kinases are found to be involved in certain signaling pathways, it usually turns out that counteracting enzymes are regulating similar processes: it has several times been reported that okadaic acid, an inhibitor of serine/threonine phosphatase PP1, interferes with integrin activation (Diamond and Springer 1994; Dumont et al. 1995; Edwards et al. 1995; Hedman and Lundgren 1996; Merrill et al. 1994).

Recently, lipid kinases and particularly phosphoinositide 3-OH kinase (PI 3-kinase) have been postulated to play significant roles in integrin activation of leukocytes and platelets. Lymphocyte cell surface receptors like CD2, CD28 and CD7 are capable of triggering α4β1 adhesion to fibronectin. These receptors are known to associate with, and activate, PI 3-kinase upon stimulation. G-protein coupled receptors also activate PI 3-kinase. Stimulation of cells with fMLP, thrombin, RANTES or MCP-1, all of which have been demonstrated to induce increases of integrin adhesiveness in the respective cell types, will induce PI 3-kinase activity (Shimizu and Hunt 1996).

Apart from these correlations, the major argument for the involvement of PI 3-kinase in integrin activation stems from the application of the relatively specific pharmacological PI 3-kinase inhibitor wortmannin. Wortmannin blocks β1-integrin adhesiveness in T cells and HL60 cells as well as the thrombin mediated αIIbβ3 activation of platelets (Shimizu et al. 1995; Zell et al. 1996; J. Zhang et al. 1996). Uncoupling the interactions of a chimeric CD28 version and PI 3-kinase in HL60 by site directed mutagenesis prevents activation of β1-integrin adhesiveness (Zell et al. 1996). These findings make a compelling but incomplete case for a role of PI-3 kinase as an important control element of integrin activation in various cell types. Future studies will have to employ activated and dominant negative mutants of PI 3-kinase in order to address its role in the control of cell adhesion more directly.

R-ras is a GTP binding protein which is highly homologous to H-ras, but it contains an additional 26 amino acids at the NH_2-terminal. Previous studies indicated that R-ras does not seem to induce cell proliferation or differentiation. Recently it was shown that expression of a constitutively active R-ras induced suspension cells to become adherent to the ECM and that this increase in adhesion results from enhanced integrin ligand binding affinity. This was demonstrated by high affinity binding of the activated integrin to monomeric fibronectin. H-ras as a control did not activate adhesion in an analogous manner while a dominant negative R-ras mutant blocked integrin activation. This suggests that R-ras plays a potent, previously unidentified, role in inside-out signaling (Z. ZHANG et al. 1996). Recent findings also implicate the small GTPase rho in the regulation of integrin affinity states (LAUDANNA et al. 1996). The mechanism is not known but may involve the actin cytoskeleton and phospholipids.

6 Proximal Regulatory Molecules at the Interface of Integrin Cytoplasmic Domains and Upstream Signaling Pathways

Although it becomes more and more evident that the activation of integrin adhesion is controlled by several upstream signal transduction pathways in various cell types, the actual proximal regulatory elements that exert their function at the level of the integrin cytoplasmic domains remained obscure. However, candidate proteins have recently been described. They were mostly identified with the help of the two-hybrid technique, i.e., by virtue of their ability to interact with the cytoplasmic portions of integrins, and for two of them there is now direct functional evidence for an involvement in the control of adhesion (Table 1).

Cytohesin-1 is a 47 kDa cytoplasmic protein which is predominantly expressed in lymphoid cells. It shares a homology region to the otherwise unrelated Sec7 gene product in yeast. Unlike Sec7, cytohesin-1 contains a COOH-terminal pleckstrin homology (PH) domain which is present in many proteins with known or postulated roles in signal transduction and which apparently may be responsible for targeting signal transducers to the plasma membrane (LEMMON et al. 1996). Cytohesin-1 was coprecipitated with $\alpha L\beta 2$ but not with $\alpha 4\beta 1$ from Jurkat cells, suggesting a direct interaction of cytohesin-1 with $\beta 2$-integrins in T cells. Overexpressed and purified

Table 1. Intracellular proteins shown to interact with integrin cytoplasmic domains directly and which have possible functions in inside-out signaling

Interacting protein	Integrin subunit	Comments
Regulatory proteins:		
cytohesin-1	b2	overexpression induces adhesion; isolated PH domain dominant negative
ILK	b1, b2, b3	overexpression suppresses adhesion
calreticulin	a2, a3, a4, av, a6	binds to GFFKR motif in a2b1 when receptor is activated
b3-endonexin	b3	binding overlaps with regulatory site in b3 cytoplasmic domain (S752)
Cytoskeletal proteins:		
a-actinin	b1, b2, b3	b2 binding enhanced in activated neutrophils
filamin	b2	binding motif overlaps with a-actinin binding site
talin	b1, b2, b3	binding sites contain NPXY motifs

cytohesin-1 from *E. coli* interacted specifically with a peptide corresponding to the complete β2 cytoplasmic domain in vitro. Functional analyses revealed that cytohesin-1 overexpression resulted in a constitutively active adhesion phenotype of αLβ2 in Jurkat cells. The expression of the isolated Sec7 domain, which mediates the interaction with β2 in yeast, had a similar although somewhat reduced effect. In marked contrast, overexpression of the PH domain resulted in a dominant negative block of β2-integrin adhesion to ICAM-1, suggesting that the PH domain, which supposedly interacts with an as yet unidentified functionally relevant ligand, plays a regulatory role in cytohesin-1 function. It is likely that the PH domain mediates membrane recruitment which points to the possibility that cytohesin-1 may indeed be directly regulated by cellular activation pathways in order to perform its function. Recently it became apparent that PH domains may bind to specific membrane bound phosphoinositol lipid ligands in vivo. The activation of lipid kinases, e.g., PI 3-kinase, may therefore result in the plasma membrane recruitment and subsequent activation of signaling proteins like cytohesin-1 via their PH domains. Most interestingly, cytohesin-1 or subdomain overexpression did not affect the adhesiveness of α4β1 which was expressed on the same cell type. Although the potential interactions of cytohesin-1 with other integrins have not been investigated yet, from these data it appears possible that cytohesin-1 is a specific regulator for β2-integrins. The mechanistic role of cytohesin-1 in avidity or affinity regulation of αLβ2 is at present unclear, but since its overexpression causes changes in the adhesive phenotype of lymphocytes directly, it is likely to become an important tool in further elucidating the activation mechanism of leukocyte integrins. Very recently it was shown that a protein which bears strong similarity to cytohesin-1 contains a GDP/GTP exchange factor domain which acts on the small G protein arf-1 in vitro. The guanine nucleotide exchange function was found to be encoded by the Sec7 domain (Chardin et al. 1996). Although direct evidence is lacking, it is tempting to speculate that cytohesin-1 may regulate cell adhesion through a specific and localized guanine nucleotide exchange function. In this context it is particularly intriguing that the isolated Sec7 domain was shown to be capable of up-regulating Jurkat cell binding to ICAM-1 (Kolanus et al. 1996).

As in the case of cytohesin-1, the integrin linked kinase (ILK) has been discovered with the help of the two-hybrid-technique. ILK is a serine/threonine specific kinase which is probably expressed ubiquitously. It was identified by its ability to bind to the β1-integrin cytoplasmic domain but apparently it can interact with other integrin β chains, too, as was demostrated by co-immunoprecipitation analyses. In vitro, ILK phosphorylates itself and exogenous substrates strongly; from these data it appears plausible to assume that the in vivo function of ILK will be dependent on the kinase activity, although this has not been demonstrated directly yet. ILK interaction with integrins may regulate both inside-out and outside-in signaling since overexpression of ILK in rat epithelial cells results in decreased cell adhesion to ECM, whereas cell spreading on fibronectin affects ILK kinase activity. Overexpression of ILK in epithelilal cells induces anchorage independent growth, stimulation of the cell cycle, and malignant transformation (Hannigan et al. 1996).

β3 endonexin interacts specifically with the cytoplasmic domain of αIIbβ3 in yeast and in vitro. This was demonstrated by binding of an endonexin fusion protein to detergent solubilized β3-integrin. The physiological role of β3 endonexin is currently unclear, but the mutation of amino acid S752P in αIIbβ3, which had previously been described to interfere with the activation of the platelet integrin, results in a marked reduction of endonexin/β3-integrin interaction in vitro (SHATTIL et al. 1995).

The calcium binding protein calreticulin binds to the membrane proximal and highly conserved GFFKR motif present within the α chain cytoplasmic domains. Deletion of this motif results in constitutive integrin adhesiveness, and an attractive hypothesis assumes the formation of a salt bridge between α and neighboring β chain residues, the disruption of which may result in integrin activation (see above). Calreticulin apparently binds to the activated but not to the inactive form of the collagen receptor, α2β1-integrin. The introduction of anti-calreticulin antibodies into Jurkat cells inhibited the ability of α2β1 to be activated by phorbol esters or by anti-integrin antibodies. Furthermore, treatment of the cells with ocadaic acid, which is known to interfere with cell adhesion, also inhibits the binding of calreticulin to the active form of α2β1(COPPOLINO et al. 1995).

7 Modulation of Integrin Adhesiveness – Potential Mechanisms

As can be seen from the above, a vast body of data has been gathered on this topic over the last 5 years or so, and plausible hypotheses which may help to explain the observed phenomena have begun to emerge.

The controversy whether the affinity or the avidity of integrins for their ligands is actually regulated by intracellular pathways is as old as the discovery of regulated adhesiveness itself and has already been discussed in excellent reviews (see Sect. 1). Another way to put this is the following: Do inside-out signaling mechanisms result in conformational changes of individual integrin molecules or do intracellular signals rather provoke the oligomerization (clustering) of the adhesion receptors without causing any direct molecular alteration of the integrins? The discussion surrounding this question continues, the main reason being that it is quite difficult to discriminate between these possibilities with any certainty in the absence of direct structural information.

"Reporter" monoclonal antibodies have been developed which recognize integrins at the cell surface only when these are in an activated state (BOUDIGNON et al. 1996; CHEN et al. 1994b; DIAMOND and SPRINGER 1994; ELEMER and EDGINGTON 1994; LUQUE et al. 1996; VAN KOOYK et al. 1991; YEDNOCK et al. 1995). The up-regulation of such activation epitopes on integrin molecules in response to cytoplasmic signals has often been interpreted as an indication that indeed conformational changes may occur within the extracellular domains which would allow enhanced ligand binding. In addition, divalent or – in the case of IgMs – multivalent binding of antibodies to their cognate antigens may also be facilitated by multimerization of integrins at the cell surface. In fact, the NK-L16 antibody, which recognizes activated αLβ2 on leukocytes, binds to the clustered β2-integrin only (Lub et al. 1995).

PAC-1 is an antibody which recognizes the activated platelet integrin exclusively. A short sequence of the CDR3 loop of PAC-1 which contains an RGD motif and which is responsible for binding of αIIbβ3 was engineered onto the antibody AP7. Binding of AP7 to αIIbβ3 is normally not much influenced by the activation state of the receptor although it recognizes the integrin with a similar but distinct element. A recombinant Fab fragment of the AP-7 derivative which bore the transplanted PAC-1 sequences behaved very much like PAC-1 (Kunicki et al. 1996). Furthermore, it was recently found that constitutively activated R-ras promotes β1-integrin binding to a monomeric fibronectin subunit (Z. Zhang et al. 1996). These data indicate that inside-out signaling events may truely aid in changing the conformations of integrin molecules. How does this occur? An attractive concept is that proximal cytoplasmic factors may bind to β or α chain cytoplasmic domains in a regulated fashion and may thereby interfere with a "salt bridge" formation between the subunits (Hughes et al. 1996). Alternatively, a currently unknown activation step may preceed binding of these factors which may then help to stabilize the unbridged state. Biochemical analyses should help to discern between these possibilities. If breaking of an integrin hinge is mediated by cytoplasmic factors like calreticulin or similar proteins just by virtue of competition for binding, then the affinities of these activation factor/cytoplasmic domain interactions should be substantially higher than the affinities between the hinge strands.

It was postulated that high affinity interactions between integrins and their ligands do not suffice to result in strong cell adhesion but that they also may require postreceptor events, which means signal transduction initiated by activated integrins. These events appeared to involve rearrangements of the membrane proximal actin cytoskeleton and the clustering of integrins at the cell surface. The mutation of a cytoplasmic element (TTT758-60AAA) of the β2 chain which was shown to be important for cell adhesion interfered with the functional association of the integrin with the cytoskeleton and thereby prevented adhesion, although the receptor was shown to be locked in a high affinity state by deletion of the GFFKR motif (Peter and O'Toole 1995). Affinity regulation and integrin signal transduction may therefore cooperate in the promotion of cell adhesiveness. Cytoplasmic regulatory molecules may play dual roles in inside-out and outside-in signaling mechanisms and such a bifunctional role was already proposed for the ILK protein (Hannigan et al. 1996).

β2-integrins bind to the ICAM-1 counter-receptor which was shown to be expressed as a noncovalently linked dimer on the cell surface (Miller et al. 1995; Reilly et al. 1995). Interestingly, leukocyte activation by phorbol esters enhances the lateral mobility of β2-integrins, probably by detachment of the integrin from the membrane proximal actin cytoskeleton (Kucik et al. 1996). An alternative or supportive mechanism may act on the actin filaments directly and has been described for platelets (Hartwig 1992): In these cells, activation events result in reduction of the average length of submembrane actin filaments which make contact to the integrin cytoplasmic domains with the help of adaptors such as α-actinin. Activation dependent severing of actin polymers appears to be mediated

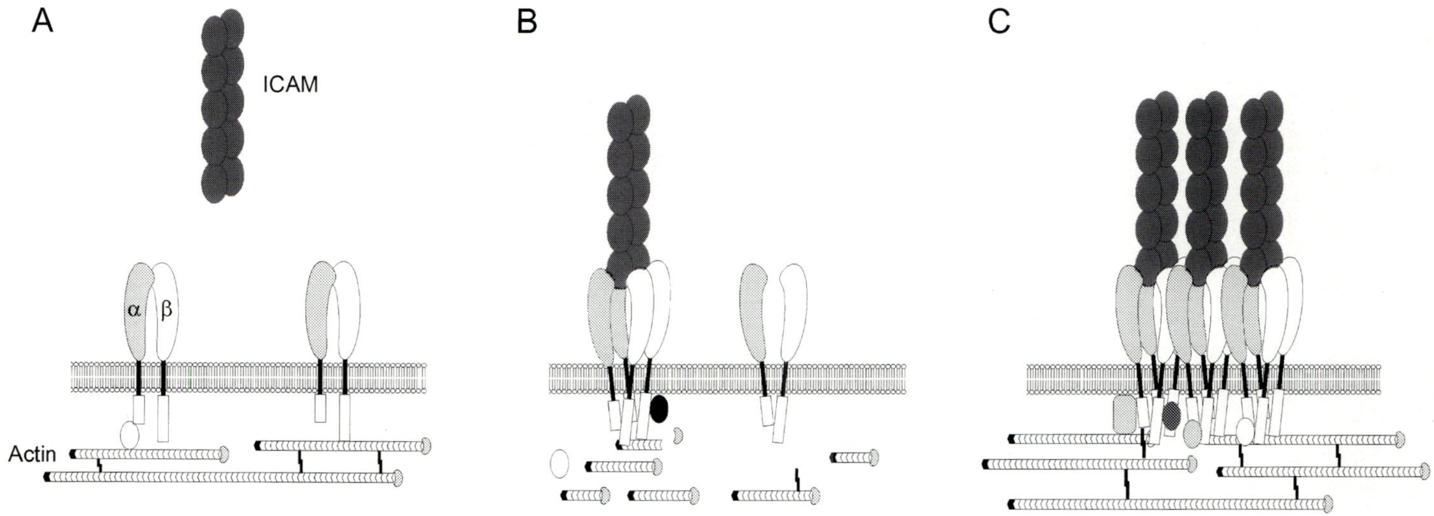

Fig. 3. Hypothetical model of integrin activation exemplified for β2 integrins. This is a modified version of a similar suggestion by Lub et al. (1995). In the resting state *A*, integrin molecules are "locked" in an inactive conformation, a process which may require cytoskeletal interactions. In an intermediately active state *B*, inside-out-mechanisms enhance lateral mobility of the integrins within the plasma membrane and may mediate their detachment from the cytoskeleton. These processes may directly lead to a high affinity state or they may facilitate and stabilize high affinity binding induced by the ligand (in this case an ICAM dimer). Subsequently, high affinity binding may lead to integrin clustering and strong adhesion *C*, a process which again appears to be dependent on the reorganization of the submembrane cytoskeleton and attachment of the integrin cytoplasmic domains to cytoskeletal proteins

by actin capping proteins, e.g., by gelsolin in platelets. It is possible that similar mechanisms hold up for leukocytes, too. Increased mobility may help the integrins to undergo low or intermediate affinity interactions with ICAM-1. The binding of ICAM-1 to $\alpha L\beta 2$ itself was proposed to induce a conformational change (high affinity binding) which may result in subsequent triggering of signal transduction, receptor clustering and tight adhesion (CABANAS and HOGG 1993). It is possible that cytoplasmic regulatory molecules play a role in all steps, disengagement from the cytoskeleton, affinity change, and postreceptor signal transduction. Since a cytoplasmic activation factor of $\alpha L\beta 2$ adhesion, cytohesin-1, has recently been discovered, it may now become feasible to dissect this highly regulated process further (KOLANUS et al. 1996) (Fig. 3).

Another aspect which requires clarification is the specificity of adhesion regulation. Often more than one type of integrin is expressed on a given cell. How does the cell ensure that only the relevant adhesion molecule becomes active in any particular physiological situation? It is much to early to answer this with any conclusiveness but cytoplasmic proteins (cytohesin-1 and β3-endonexin) which couple to their target elements (β2- and β3-integrins, respectively) with at least some specificity have now been described (KOLANUS et al. 1996; SHATTIL et al. 1995). Furthermore, the dissection of apparently independent upstream signaling pathways which lead to either the activation of β2- or of β1-integrins on the same cell has recently been reported (CARR et al. 1996; WEBER et al. 1996; LAUDANNA et al. 1996). From these data it appears rather plausible that such specificities do indeed exist.

Inside-out signal transduction of integrins has been demonstrated to be a significant pathway in the dynamic regulation of cell adhesion. Proximal and distal intracellular control elements which play important roles in these processes have begun to emerge. Future research efforts will broaden our understanding of the involved mechanisms and the resulting informations may also be used to target key regulatory molecules with pharmacological inhibitors. Such an interference with signaling processes that control cell adhesion may be useful as an effective and specific means of mediating immunosuppression in treatment of diseases, e.g., chronic inflammation and sepsis, or in the counteraction of graft rejection in organ transplantation.

References

Boudignon PC, Patel PM, Parise LV (1996) Phorbol ester enhances integrin a IIb b 3-dependent adhesion of human erythroleukemic cells to activation-dependent monoclonal antibodies. Blood 87:968–976

Cabanas C, Hogg N (1993) Ligand intercellular adhesion molecule 1 has a necessary role in activation of integrin lymphocyte function-associated molecule 1. Proc Natl Acad Sci USA 90:5838–5842

Carr MW, Alon R, Springer TA (1996) The C-C chemokine MCP-1 differentially modulates the avidity of b 1 and b 2 integrins on T lymphocytes. Immunity 4:179–187

Chan BM, Kassner PD, Schiro JA, Byers HR, Kupper TS, Hemler ME (1992) Distinct cellular functions mediated by different VLA integrin a subunit cytoplasmic domains. Cell 68:1051–1060

Chardin P, Paris S, Antonny B, Robineau S, Beraud-Dufour S, Jackson CL, Chabre M (1996) A human exchange factor for Arf contains Sec-7 and pleckstrin homology domains. Nature 384:481–484

Chen YP, Djaffar I, Pidard D, Steiner B, Cieutat AM, Caen JP, Rosa JP (1992) Ser-752 → Pro mutation in the cytoplasmic domain of integrin b 3 subunit and defective activation of platelet integrin a IIb b 3 (glycoprotein IIbIIIa) in a variant of Glanzmann thrombasthenia. Proc Natl Acad Sci USA 89: 10169–10173

Chen YP, O'Toole TE, Shipley T, Forsyth J, LaFlamme SE, Yamada KM, Shattil SJ, Ginsberg MH (1994a) "Inside-out" signal transduction inhibited by isolated integrin cytoplasmic domains. J Biol Chem 269:18307–18310

Chen YP, O'Toole TE, Ylanne J, Rosa JP, Ginsberg MH (1994b) A point mutation in the integrin b 3 cytoplasmic domain (S752 → P) impairs bidirectional signaling through a IIb b 3 (platelet glycoprotein IIbIIIa). Blood 84:1857–1865

Chun JS, Ha MJ, Jacobson BS (1996) Differential translocation of protein kinase C epsilon during HeLa cell adhesion to a gelatin substratum. J Biol Chem 271:13008–13012

Coppolino M, Leung HC, Dedhar S, Wilkins J (1995) Inducible interaction of integrin a 2 b 1 with calreticulin. Dependence on the activation state of the integrin. J Biol Chem 270:23132–23138

Crowe DT, Chiu H, Fong S, Weissman IL (1994) Regulation of the avidity of integrin alpha 4 b 7 by the b 7 cytoplasmic domain. J Biol Chem 269:14411–14418

Dedhar S, Hannigan GE (1996) Integrin cytoplasmic interactions and bidirectional transmembrane signalling. Curr Opin Cell Biol 8:657–669

Diamond MS, Springer TA (1994) The dynamic regulation of integrin adhesiveness. Curr Biol 4:506–517

Dumont FJ, Fischer P, Sirotina A (1995) Increased LFA-1-mediated homotypic cell adhesion is associated with the G1 growth arrest induced by rapamycin in a T cell lymphoma. Exp Cell Res 219: 146–158

Dustin ML, Springer TA (1989) T cell receptor cross-linking transiently stimulates adhesiveness through LFA-1. Nature 341:619–624

Edwards BS, Curry MS, Southon EA, Chong AS, Graf LJ (1995) Evidence for a dithiol-activated signaling pathway in natural killer cell avidity regulation of leukocyte function antigen-1: structural requirements and relationship to phorbol ester- and CD16-triggered pathways. Blood 86:2288–2301

Eguchi H, Horikoshi T (1996) The expression of integrin a 2 b 1 and attachment to type I collagen of melanoma cells are preferentially induced by tumour promoter, TPA (12-O-tetradecanoyl phorbol-13-acetate). Br J Dermatol 134:33–39

Elemer GS, Edgington TS (1994) Monoclonal antibody to an activation neoepitope of a M b 2 inhibits multiple a M b 2 functions. J Immunol 152:5836–5844

Filardo EJ, Cheresh DA (1994) A b turn in the cytoplasmic tail of the integrin a v subunit influences conformation and ligand binding of a v b 3. J Biol Chem 269:4641–4647

Gabbeta J, Yang X, Sun L, McLane MA, Niewiarowski S, Rao AK (1996) Abnormal inside-out signal transduction-dependent activation of glycoprotein IIbIIIa in a patient with impaired pleckstrin phosphorylation. Blood 87:1368–1376

Ginsberg MH, Du X, Plow EF (1992) Inside-out integrin signalling. Curr Opin Cell Biol 4:766–771

Hannigan GE, Leung HC, Fitz GL, Coppolino MG, Radeva G, Filmus J, Bell JC, Dedhar S (1996) Regulation of cell adhesion and anchorage-dependent growth by a new b 1-integrin-linked protein kinase. Nature 379:91–96

Hartwig JH (1992) Mechanisms of actin rearrangements mediating platelet activation. J Cell Biol 118:1421–1442

Hauss P, Mazerolles F, Hivroz C, Lecomte O, Barbat C, Fischer A (1993) GF109203X, a specific PKC inhibitor, abrogates anti-CD3 antibody-induced upregulation of CD4+ T cell adhesion to B cells. Cell Immunol 150:439–446

Hayashi Y, Haimovich B, Reszka A, Boettiger D, Horwitz A (1990) Expression and function of chicken integrin b 1 subunit and its cytoplasmic domain mutants in mouse NIH 3T3 cells. J Cell Biol 110: 175–184

Hedman H, Lundgren E (1996) Regulation of a 4 integrin avidity in human B cells: requirement for dephosphorylation events for high avidity VCAM-1 binding. Scand J Immunol 44:239–242

Hemler ME (1990) VLA proteins in the integrin family: structures, functions, and their role on leukocytes. Annu Rev Immunol 8:365–400

Hermanowski VA, Van SJ, Swiggard WJ, Wright SD (1992) Integrin modulating factor-1: a lipid that alters the function of leukocyte integrins. Cell 68:341–352

Hibbs ML, Jakes S, Stacker SA, Wallace RW, Springer TA (1991a) The cytoplasmic domain of the integrin lymphocyte function-associated antigen 1 b subunit: sites required for binding to intercellular adhesion molecule 1 and the phorbol ester-stimulated phosphorylation site. J Exp Med 174: 1227–1238

Hibbs ML, Xu H, Stacker SA, Springer TA (1991b) Regulation of adhesion of ICAM-1 by the cytoplasmic domain of LFA-1 integrin b subunit. Science 251:1611–1613

Hirsch E, Iglesias A, Potocnik AJ, Hartmann U, Fassler R (1996) Impaired migration but not differentiation of haematopoietic stem cells in the absence of b1 integrins. Nature 380:171–175

Horwitz AF, Duggan K, Buck C, Beckerle MC, Burridge K (1986) Interactions of plasma membrane fibronectin receptor with talin: a transmembrane linkage. Nature 320:531–540

Hughes PE, O'Toole TE, Ylanne J, Shattil SJ, Ginsberg MH (1995) The conserved membrane-proximal region of an integrin cytoplasmic domain specifies ligand binding affinity. J Biol Chem 270: 12411–12417

Hughes PE, Diaz GF, Leong L, Wu C, McDonald JA, Shattil SJ, Ginsberg MH (1996) Breaking the integrin hinge A defined structural constraint regulates integrin signaling. J Biol Chem 271:6571–6574

Hynes RO (1992) Integrins: versatility, modulation, and signaling in cell adhesion. Cell 69:11–25

Kassner PD, Hemler ME (1993) Interchangeable a chain cytoplasmic domains play a positive role in control of cell adhesion mediated by VLA-4, a b 1 integrin. J Exp Med 178:649–660

Kassner PD, Kawaguchi S, Hemler ME (1994) Minimum a chain cytoplasmic tail sequence needed to support integrin-mediated adhesion. J Biol Chem 269:19859–19867

Kawaguchi S, Hemler ME (1993) Role of the a subunit cytoplasmic domain in regulation of adhesive activity mediated by the integrin VLA-2. J Biol Chem 268:16279–16285

Kiley SC, Parker PJ (1995) Differential localization of protein kinase C isozymes in U937 cells: evidence for distinct isozyme functions during monocyte differentiation. J Cell Sci 108:1003–1016

Klemke RL, Yebra M, Bayna EM, Cheresh DA (1994) Receptor tyrosine kinase signaling required for integrin a v b 5-directed cell motility but not adhesion on vitronectin. J Cell Biol 127:859–866

Kolanus W, Nagel W, Schiller B, Zeitlmann L, Godar S, Stockinger H, Seed B (1996) A L b 2 integrin/LFA-1 binding to ICAM-1 induced by cytohesin-1, a cytoplasmic regulatory molecule. Cell 86: 233–242

Kucik DF, Dustin ML, Miller JM, Brown EJ (1996) Adhesion-activating phorbol ester increases the mobility of leukocyte integrin LFA-1 in cultured lymphocytes. J Clin Invest 97:2139–2144

Kunicki TJ, Annis DS, Deng YJ, Loftus JC, Shattil SJ (1996) A molecular basis of affinity modulation of Fab ligand binding to integrin alphaIIb b 3. J Biol Chem 271:20315–20321

Laudanna C, Campbell JJ, Butcher EC (1996) Role of Rho in chemoattractant-activated leukocyte adhesion through integrins. Science 271:981–983

Lee DH, MacIntyre JP, Wang E, Hudson DJ, Ishaque A, Conant JA, Pope BL, Lau CY (1994) A leukocyte lipid up-regulates the avidity of lymphocyte function-associated antigen-1. Biochem Biophys Res Commun 199:319–326

Lemmon MA, Ferguson KM, Schlessinger J (1996) PH domains: diverse sequences with a common fold recruit signaling molecules to the cell surface. Cell 85:621–624

Lewis JM, Schwartz MA (1995) Mapping in vivo associations of cytoplasmic proteins with integrin b 1 cytoplasmic domain mutants. Mol Biol Cell 6:151–160

Loh E, Beaverson K, Vilaire G, Qi W, Poncz M, Bennett JS (1995) Agonist-stimulated ligand binding by the platelet integrin a IIb b 3 in a lymphocyte expression system. J Biol Chem 270:18631–18636

Lub M, van Kooyk Y, Figdor CF (1995) Ins and outs of LFA-1. Immunol Today 16:479–483

Luque A, Gomez M, Puzon W, Takada Y, Sanchez MF, Cabanas C (1996) Activated conformations of very late activation integrins detected by a group of antibodies (HUTS) specific for a novel regulatory region (355–425) of the common b 1 chain. J Biol Chem 271:11067–11075

Merrill JT, Winchester RJ, Buyon JP (1994) Dynamic state of b 2 integrin phosphorylation: regulation of neutrophil aggregation involves a phosphatase-dependent pathway. Clin Immunol Immunopathol 71:216–222

Miller J, Knorr R, Ferrone M, Houdei R, Carron CP, Dustin ML (1995) Intercellular adhesion molecule-1 dimerization and its consequences for adhesion mediated by lymphocyte function associated-1. J Exp Med 182:1231–1241

Mobley JL, Ennis E, Shimizu Y (1996) Isolation and characterization of cell lines with genetically distinct mutations downstream of protein kinase C that result in defective activation-dependent regulation of T cell integrin function. J Immunol 156:948–956

O'Toole TE, Mandelman D, Forsyth J, Shattil SJ, Plow EF, Ginsberg MH (1991) Modulation of the affinity of integrin a IIb b 3 (GPIIbIIIa) by the cytoplasmic domain of a IIb. Science 254:845–847

O'Toole TE, Katagiri Y, Faull RJ, Peter K, Tamura R, Quaranta V, Loftus JC, Shattil SJ, Ginsberg MH (1994) Integrin cytoplasmic domains mediate inside-out signal transduction. J Cell Biol 124: 1047–1059

O'Toole TE, Ylanne J, Culley BM (1995) Regulation of integrin affinity states through an NPXY motif in the b subunit cytoplasmic domain. J Biol Chem 270:8553–8558

Otey CA, Pavalko FM, Burridge K (1990) An interaction between a-actinin and the b 1 integrin subunit in vitro. J Cell Biol 111:721–729

Otey CA, Vasquez GB, Burridge K, Erickson BW (1993) Mapping of the a-actinin binding-site within the B-1 integrin cytoplasmic domain. J Biol Chem 268:21193–21197

Pacifici R, Roman J, Kimble R, Civitelli R, Brownfield CM, Bizzarri C (1994) Ligand binding to monocyte a 5 b 1 integrin activates the a 2 b 1 receptor via the a 5 subunit cytoplasmic domain and protein kinase C. J Immunol 153:2222–2233

Pasqualini R, Hemler ME (1994) Contrasting roles for integrin b 1 and b 5 cytoplasmic domains in subcellular localization, cell proliferation, and cell migration. J Cell Biol 125:447–460

Pavalko FM, Laroche SM (1993) Activation of human neutrophils induces an interaction between the integrin-B-2-subunit (CD18) and the actin-binding protein A-actinin. J Immunol 151:3795–3807

Peter K, O'Toole TE (1995) Modulation of cell adhesion by changes in a L b 2 (LFA-1, CD11a/CD18) cytoplasmic domain/cytoskeleton interaction. J Exp Med 181:315–326

Reilly PL, Woska JJ, Jeanfavre DD, McNally E, Rothlein R, Bormann BJ (1995) The native structure of intercellular adhesion molecule-1 (ICAM-1) is a dimer Correlation with binding to LFA-1. J Immunol 155:529–532

Sharma CP, Ezzell RM, Arnaout MA (1995) Direct interaction of filamin (ABP-280) with the b 2-integrin subunit CD18. J Immunol 154:3461–3470

Shattil SJ, O'Toole T, Eigenthaler M, Thon V, Williams M, Babior BM, Ginsberg MH (1995) B 3-endonexin, a novel polypeptide that interacts specifically with the cytoplasmic tail of the integrin b 3 subunit. J Cell Biol 131:807–816

Shimizu Y, Hunt SW (1996) Regulating integrin-mediated adhesion: one more function for PI 3-kinase? Immunol Today 17:565–573

Shimizu Y, Mobley JL, Finkelstein LD, Chan AS (1995) A role for phosphatidylinositol 3-kinase in the regulation of b 1 integrin activity by the CD2 antigen. J Cell Biol 131:1867–1880

Springer TA (1990) Adhesion receptors of the immune system. Nature 346:425–434

Valmu L, Gahmberg CG (1995) Treatment with okadaic acid reveals strong threonine phosphorylation of CD18 after activation of CD11/CD18 leukocyte integrins with phorbol esters or CD3 antibodies. J Immunol 155:1175–1183

van Kooyk Y, van de Wiel, van Kemenade P, Weder P, Kuijpers TW, Figdor CG (1989) Enhancement of LFA-1-mediated cell adhesion by triggering through CD2 or CD3 on T lymphocytes. Nature 342:811–813

van Kooyk Y, Weder P, Hogervorst F, Verhoeven AJ, van Seventer G, te Velde AA, Borst J, Keizer GD, Figdor CG (1991) Activation of LFA-1 through a Ca2(+)-dependent epitope stimulates lymphocyte adhesion. J Cell Biol 112:345–354

van Leeuwen R, Dekker SK, Arbiser JL, Vermeer BJ, Bruijn JA, Byers HR (1994) Phorbol ester induced rapid attachment and spreading of melanoma cells and the role of extracellular matrix proteins. Int J Cancer 57:894–900

van Willigen G, Hers I, Gorter G, Akkerman JW (1996) Exposure of ligand-binding sites on platelet integrin a IIB/b 3 by phosphorylation of the b 3 subunit. Biochem J 314:769–779

Weber C, Katayama J, Springer TA (1996) Differential regulation of b 1 and b 2 integrin avidity by chemoattractants in eosinophils. Proc Natl Acad Sci USA 93:10939–10944

Wu C, Keivens VM, O'Toole TE, McDonald JA, Ginsberg MH (1995) Integrin activation and cyto-skeletal interaction are essential for the assembly of a fibronectin matrix. Cell 83:715–724

Yang JT, Rayburn H, Hynes RO (1993) Embryonic mesodermal defects in a 5 integrin-deficient mice. Development 119:1093–1105

Yang JT, Rayburn H, Hynes RO (1995) Cell adhesion events mediated by a 4 integrins are essential in placental and cardiac development. Development 121:549–560

Yednock TA, Cannon C, Vandevert C, Goldbach EG, Shaw G, Ellis DK, Liaw C, Fritz LC, Tanner LI (1995) A 4 b 1 integrin-dependent cell adhesion is regulated by a low affinity receptor pool that is conformationally responsive to ligand. J Biol Chem 270:28740–28750

Zell T, Hunt SR, Mobley JL, Finkelstein LD, Shimizu Y (1996) CD28-mediated up-regulation of b1-integrin adhesion involves phosphatidylinositol 3-kinase. J Immunol 156:883–886

Zhang J, Zhang J, Shattil SJ, Cunningham MC, Rittenhouse SE (1996) Phosphoinositide 3-kinase gamma and p85/phosphoinositide 3-kinase in platelets. Relative activation by thrombin receptor or b-phorbol myristate acetate and roles in promoting the ligand-binding function of aIIbb3 integrin. J Biol Chem 271:6265–6272

Zhang Z, Vuori K, Wang H, Reed JC, Ruoslahti E (1996) Integrin activation by R-ras. Cell 85:61–69

The Role of α4-Integrin in T Lymphocyte Migration Into the Inflamed and Noninflamed Central Nervous System

B. ENGELHARDT

1 Introduction . . . 51

2 The Blood-Brain Barrier . . . 52

3 Models of T Cell Immigration Into the CNS . . . 52
3.1 Immigration of T Cells Into the Healthy CNS . . . 53
3.2 Immigration of T Lymphocytes Into the Inflamed CNS . . . 54

4 Molecular Mechanisms of T Cell Migration Across the Blood-Brain Barrier . . . 54
4.1 The Multistep Paradigm . . . 54
4.2 Expression of CAMs at the Blood-Brain Barrier . . . 55
4.3 Phenotype of CNS-Seeking T Cells . . . 56
4.4 Functional Role of CAMs in T Cell – Blood-Brain Barrier Interaction . . . 57
4.4.1 In Vitro . . . 57
4.4.2 In Vivo . . . 58

5 The Key Role of α4-Integrin . . . 58

6 Conclusion . . . 61

References . . . 62

1 Introduction

The central nervous system (CNS) is considered to be an immunologically privileged site. The CNS parenchyme is tightly sealed off from the periphery by the highly specialized vascular wall – the blood-brain barrier (BBB). The BBB has been assumed to provide a barrier that prevents circulating leukocytes from entering the CNS. However, in inflammatory reactions in response to virus infections, in multiple sclerosis and in experimental autoimmune encephalomyelitis (EAE), mononuclear cells infiltrate the CNS. To date, there is no evidence that the functions of the BBB endothelium include BBB-specific leukocyte/endothelial interactions that regulate leukocyte recruitment across the BBB. Therefore it is thought that the interaction of circulating "CNS-seeking" mononuclear cells with the endothelial cells of the BBB is a multistep process (BUTCHER 1991; SPRINGER 1994; BARGATZE

Max-Planck-Institut für physiologische und klinische Forschung, W.G. Kerckhoff-Institut, 61231 Bad Nauheim, Germany et al. 1995). This review summarizes the current evidence, pointing to a unique role of α4-integrin in the interaction of circulating mononuclear cells with the endothelium of the BBB.

2 The Blood-Brain Barrier

The BBB separates the CNS tissue from the blood by preventing the free exchange of molecules between the blood and neuropil. Fully differentiated brain endothelial cells build a physical barrier for hydrophilic substances. The endothelial cells of the brain lack pinocytotic vesicles and thus do not carry out transcellular bulk flux. Additionally, there is a highly specialized and complex network of tight junctions between the endothelial cells which prevents paracellular diffusion (NAGY et al. 1984; NICO et al. 1992; WOLBURG et al. 1994). Besides its physical barrier characteristics, the BBB provides a metabolic barrier by the expression of specific cytoplasmic enzymes and by the polar expression of specific transport systems at either the luminal or abluminal surfaces of the brain endothelium (for review see ENGELHARDT and RISAU 1995). Thus the BBB ensures the transport of sufficient metabolites from the blood into the brain and quick removal of toxic substances from the brain. While the endothelial cells form the barrier proper, the permanent complex interactions of endothelial cells with adjacent cells like pericytes, perivascular microglial cells and astrocytes seem to be a prerequisite for the maintenance of barrier function.

3 Models of T Cell Immigration Into the CNS

There is increasing evidence that immunological mechanisms are involved in a considerable number of disease processes of the CNS. Viruses from several families can infect the CNS causing distinct diseases depending in part on the type of CNS cell infected (reviewed by GRIFFIN et al. 1992). The most severe infections are caused by viruses such as herpes simplex type I, which replicates primarily in neurons. There are animal models for most of these diseases, allowing the study of pathogenesis of the disease (reviewed by GRIFFIN et al. 1992). In these models, virus is usually injected intrathecally in susceptible animal strains. Virus replication within the CNS then causes recruitment of mononuclear cells into the CNS. Perivascular cuffs of mononuclear cells around small arteries, arterioles and venules are the classic hallmark of viral infection of the CNS. There is also accumulating evidence for a pathogenic role for T cells immigrating into the CNS in some models of virus-induced encephalomyelitis (TSUNODA and FUJINAMI 1996; MUSETTE et al. 1995; LIEBERT and TER MEULEN

1993; RICHT et al. 1989). The molecular mechanisms mediating the recruitment of T lymphocytes across the BBB in virus-induced encephalomyelitis have not been investigated in detail to date (IRANI and GRIFFIN 1996). The same holds true for encephalomyelitis caused by neurotropic parasites, such as *Toxoplasma gondii*, which enter the CNS and cause a severe inflammatory reaction (DECKERT-SCHLÜTER et al. 1994). Injection of attenuated bacteria into the CNS of sensitized mice will elicit recruitment of mononuclear cells consisting mostly of T cells into the CNS; however, the number of recruited mononuclear cells is low (ENGELHARDT et al. 1995). There have also been efforts to induce the recruitment of inflammatory cells into the CNS by intracerebral injection of cytokines such as lipopolysaccharide (LPS), tumor necrosis factor (TNF-α), or interleukin (IL)-1 (ANDERSSON et al. 1992a, b). Injection of cytokines has been shown to induce marginal leukocyte extravasation across the BBB but not emigration of mononuclear cells into the CNS parenchyme.

Investigations into the molecular mechanisms involved in T cell recruitment into the CNS in vivo are still few. Most studies focused on the expression of adhesion molecules but rarely on their functional importance for cell recruitment into the inflamed CNS (IRANI and GRIFFIN 1996). Our current knowledge regarding the molecular mechanisms involved in T cell immigration into the CNS is mainly limited to studies performed in the animal model experimental autoimmune encephalomyelitis (EAE). EAE is a T cell-mediated autoimmune disease of the CNS, that has clinical and histopathological characteristics which make it the prototype animal model for human inflammatory demyelinating diseases of the CNS, e.g., multiple sclerosis (MS; MARTIN and MCFARLAND 1995). EAE is mediated by autoantigen-specific $CD4^+$ T helper-1 cells. It can be induced in susceptible animal strains by immunization with spinal cord homogenate or purified myelin antigens. The two major proteins of the CNS myelin that induce EAE are myelin basic protein (MBP) and protein lipid protein (PLP) (CHOU et al. 1983; LEES et al. 1989). Also, EAE can be adoptively transferred by intravenous injection of autoantigen-specific T cells into naive syngeneic recipients.

3.1 Immigration of T Cells Into the Healthy CNS

Under normal circumstances the brain does not seem to be surveyed by circulating lymphocytes to the same degree as extracerebral tissues (WEKERLE et al. 1986). The fact that intravenously injected T cells, however, are able to induce EAE and an inflammatory response in the CNS provided evidence that the normal BBB is not an impermeable barrier for all immune cells. The first observations that activated, antigen-specific T line cells are capable of penetrating the healthy BBB were made by WEKERLE et al. (1986) and extended by HICKEY and coworkers (1991). Both groups traced the migration pattern of intravenously injected [^{14}C]thymidine $CD4^+$ MBP-specific T line cells in Lewis rats and demonstrated labeled cells within the CNS 6 h after injection. It should be noted that both studies failed to show immigration of resting T line cells into the CNS. Failure of resting T cells to penetrate the BBB might therefore offer an explanation for the observation that activated but not resting autoaggressive T cells can be used to transfer EAE to syngeneic recipients.

3.2 Immigration of T Lymphocytes Into the Inflamed CNS

An initial insult – whether infection, autoimmune, or direct application of proinflammatory stimulants – leads to local tissue damage and production of cytokines. These in turn cause stimulation of the BBB endothelium and thus an altered phenotype of the BBB. During the acute phase of these inflammatory responses within the CNS there is evidence for BBB leakiness, as documented by the presence of serum albumin in the perivascular space. Nevertheless, there is no evidence for passive "leakage" of cellular elements across the inflamed BBB, as documented by the absence of erythrocytes in the inflammatory infiltrate. Immigration of mononuclear cells across the stimulated BBB rather seems to be a selective process, as the inflammatory infiltrates in viral- or parasite-induced encephalitis, as well as during EAE, can be characterized by the presence of certain subsets of mononuclear cells and the absence of others (RENNO et al. 1994; STEFFEN et al. 1994; DECKERT-SCHLÜTER et al. 1994; ENGELHARDT et al. 1995; IRANI and GRIFFIN 1996). It should be noted that not even the induction of inflammatory reactions by direct intracerebral injection of proinflammatory stimuli results in random immigration of leukocytes into the CNS (ANDERSSON et al. 1992a, b; ENGELHARDT et al. 1995). Thus although there is evidence for BBB leakage during CNS inflammation, the mechanisms for immune cell entry across the stimulated BBB into the CNS do not seem to be random at any time.

4 Molecular Mechanisms of T Cell Migration Across the Blood-Brain Barrier

4.1 The Multistep Paradigm

Leukocyte extravasation into peripheral tissues is exquisitely regulated in vivo by mechanisms of selective leukocyte-endothelial cell recognition, which can display extraordinary specificity depending on the tissue site or organ involved, the nature of an inflammatory stimulus and also the timepoint during the inflammatory response. The observation that the specificity of leukocyte recruitment cannot be explained by simple lock-and-key models of leukocyte-endothelial interaction led BUTCHER (1991) to propose a model which views leukocyte-endothelial interaction as an active process requiring multiple steps. The multistep paradigm postulates that traffic signals for leukocytes function in a sequence of steps, allowing multiple molecular choices at each step, thus providing great combinatorial diversity in signaling. The selectins have been shown to regulate the first step of this process, the tethering of leukocytes from the flowing blood to the vascular wall. The subsequent firm adhesion of leukocytes to the endothelial cell surface and migration across the vessel wall (diapedesis) are regulated independently of the initial binding. These latter stages are thought to be mediated by functional activation of integrins on the leukocyte by chemoattractants located in the vessel wall (BUTCHER 1991; SPRINGER 1994). The activated integrins mediate firm adhesion of the leukocyte to the endothelium by interaction with their endothelial ligands that are members of the immunoglobulin superfamily. The multistep paradigm has been shown to hold true for neutrophil recruitment across vascular walls. Also there is accumulating evidence that recruitment of monocytes and lymphocytes follows the same paradigm (BARGATZE et al. 1995).

4.2 Expression of Cell Adhesion Molecules at the Blood-Brain Barrier

With respect to the structural uniqueness of the BBB endothelium, the major question regarding T cell migration into the CNS has been whether the differentiation of BBB endothelium extends to the presence of BBB-specific cell adhesion molecules (CAMs). To date, there is no evidence for the induction of BBB specific CAMs, although the expression of CAMs on the BBB during the progression of inflammatory conditions of the CNS has been the subject of extensive research (CANNELLA et al. 1990, 1991a, b; LASSMANN et al. 1991; SASSEVILLE et al. 1992; DECKERT-SCHLÜTER et al. 1994; ENGELHARDT et al. 1994; STEFFEN et al. 1994; IRANI and GRIFFIN 1996).

With regard to the expression of CAMs, cerebral endothelial cells in the healthy CNS do not differ from extracerebral endothelial cells, with the exception of lower expression of vascular endothelial (VE)-cadherin in cerebral endothelium than in extracerebral endothelial cells (BREIER et al. 1995). In the rat and in humans, expression of LFA-3 has been reported on the BBB endothelium (JUNG et al. 1995; RÖSSLER et al. 1992). There are low levels of constitutive expression of ICAM-1, and VCAM-1, on some larger venules within the healthy CNS (ENGELHARDT et al. 1994; IRANI and GRIFFIN 1996; STEFFEN et al. 1994). Massive up-regulation of ICAM-1 and VCAM-1 on CNS microvasculature has been reported in a wide array of inflammatory conditions including virus- and parasite-induced encephalitis (IRANI and GRIFFIN 1996; SASSEVILLE et al. 1992; DECKERT-SCHLÜTER et al. 1994) in response to tissue injury and stimulation by attenuated bacteria (ENGELHARDT et al. 1994) and in EAE. In EAE up-regulation of ICAM-1 and VCAM-1 precedes the perivascular inflammatory cell cuffing and the onset of clinical disease (CANNELLA et al. 1990; Schulz and Engelhardt, unpublished). Expression of ICAM-1 and VCAM-1 remains high throughout the first clinical episode (STEFFEN et al. 1994). Modulation of endothelial ICAM-1 expression in the CNS correlates with the clinical time course of chronically relapsing remitting EAE; up-regulated levels of ICAM-1 were shown to drop after the first clinical episode of EAE and be up-regulated again prior to the clinical relapse and development of new inflammatory infiltrates (CANNELLA et al. 1990). The observation that up-regulation of ICAM-1 and VCAM-1 on cerebral vessels is not restricted to vessels surrounded by an inflammatory infiltrate further suggests that expression of these molecules is necessary but not sufficient for inflammatory cell entry into the CNS.

During chronic inflammation further changes of cerebral vessels have been noted leading to a morphology which resembles high endothelial venules (HEVs) in lymph nodes (RAINE et al. 1990). The induction of MAdCAM-1 has been reported on such HEV-like cerebral vessels during chronic EAE in the Biozzi-mouse strain (O'NEILL et al. 1991). However, MAdCAM-1 could not be seen in chronic inflammation during EAE in the SJL/J mouse strain (Schulz and Engelhardt, unpublished). Also, there has been one report describing reactivity of inflamed vessels with the monoclonal antibody HECA-452 in multiple sclerosis brains (RAINE et al. 1990). HECA-452 recognizes the E-selectin ligand cutaneous lymphocyte antigen (CLA) on a subpopulation of T memory cells (PICKER et al. 1993), but also a carbohydrate moiety on HEVs in mice. Other selectin ligands have not been demonstrated on cerebral vessels during inflammation to date. Furthermore, it has been shown that brain endothelial cells lack storage of P-selectin in their Weibel-Palade bodies (BARKALOW et al. 1996). In addition we have shown that E- and P-selectin are not induced on cerebral vessels during EAE as their inducibility is suppressed by the CNS microenvironment in vivo (ENGELHARDT et al., 1997). E-selectin expression has, however, been demonstrated on some vessels from multiple sclerosis brains (WASHINGTON et al. 1994). Thus, E-selectin expression on cerebral vessels might occur in very late stages of ongoing CNS inflammation, after massive destruction of the surrounding tissue.

4.3 Phenotype of CNS-Seeking T Cells

Several studies characterized the phenotype of T lymphocytes infiltrating the CNS during different inflammatory conditions using immunohistochemistry or by performing FACS analysis on inflammatory cells isolated from the inflamed CNS tissue. It seems that independent of the inflammatory cause, CNS infiltrating T cells are recently activated/memory lymphocytes, as characterized by their high level expression of CD44, LFA-1 and ICAM-1 and low level expression of CD45RB in the mouse (BARTEN et al. 1995; ENGELHARDT et al. 1995; RENNO et al. 1994; HURWITZ et al. 1992; ZEINE and OWENS 1992). This by itself does not distinguish inflammatory T cells in the CNS from the inflammatory T cell infiltrating other nonlymphoid tissues (MACKAY 1991). However, inflammatory T cells in the CNS differ from T cells present in extracerebral inflammatory sites by their distinct expression of certain integrins (ENGELHARDT et al. 1995; ENGELHARDT et al., submitted). Inflammatory T cells in the CNS lack expression of α6-integrin and the αE-integrin chain (ENGELHARDT et al. 1995; ENGELHARDT et al., submitted) and express low to undetectable levels of the $\alpha4\beta7$-heterodimer (ENGELHARDT et al. 1995). However, they do express α4-integrin at the same level as circulating T cells implying that inflammatory T cells present in the CNS only express the $\alpha4\beta1$-heterodimer. Recent studies in our laboratory indicate that T cells in the CNS during EAE in the SJL/J mouse in fact do express α4β1 in a highly activated state as demonstrated by positive staining with a monoclonal antibody specifically detecting activated β1 in context with α4-integrin (LENTER et al. 1993; ENGELHARDT et al., submitted). A subpopulation of T cells present in the CNS during EAE expresses high affinity ligands for E- and P-selectin (ENGELHARDT et al., 1997). However, expression of L-selectin by these cells is still controversial (DOPP et al. 1994; ENGELHARDT et al. 1995). Due to the fact that L-selectin can be down-regulated upon lymphocyte contact with endothelium its expression on CNS-seeking T cells remains speculative.

4.4 Functional Role of CAMs in T Cell – Blood-Brain Barrier Interaction

4.4.1 In Vitro

The expression of CAMs, such as ICAM-1 and VCAM-1, on the BBB does not allow any conclusions regarding their functional importance in T cell interaction, as their integrin ligands are activation dependent adhesion receptors. Therefore, evidence for the functional importance of cerebral endothelial ICAM-1 and VCAM-1 has been provided by several studies using a modified version of the Stamper-Woodruff frozen section adhesion assay (STAMPER and WOODRUFF 1976). In these assays, frozen sections derived from brain tissue of animals afflicted with EAE (STEFFEN et al. 1994; YEDNOCK et al. 1992) or virus-induced encephalitis (SASSEVILLE et al. 1994) were coincubated with immune cells. These studies clearly demonstrate that mononuclear cells can bind to inflamed cerebral vessels with increased levels of ICAM-1 and VCAM-1 via the interaction of their known ligands LFA-1/Mac-1 and α4-integrins. With respect to the lack of α4β7-integrin expressing T lymphocytes within the inflamed CNS it should be noted that T cells, which have been shown to express highly activated α4β7-integrins, can bind to VCAM-1 expressed on inflamed cerebral endothelium via α4β7. However, binding of lymphocytes derived from inflamed peripheral lymph nodes could clearly be shown to be mediated via LFA-1/ICAM-1 and α4β1/VCAM-1 (STEFFEN et al. 1994). Binding of lymphocytes to vessels in healthy brains could not be demonstrated in this assay.

Investigation of the binding characteristics of inflammatory cells isolated from inflamed brains on purified ICAM-1, VCAM-1, and MAdCAM-1 provided information about their possible usage of certain integrin receptors. Lymphocytes isolated from inflamed brains could be shown to bind to ICAM-1 and VCAM-1, but rarely to MAdCAM-1 indicating that they actively use LFA-1 and α4β1 integrins for binding to endothelium (ENGELHARDT et al. 1995).

Taken together, the in vitro data provided evidence for an involvement of ICAM-1 and VCAM-1 in lymphocyte interactions with cerebral endothelium during inflammation.

4.4.2 In Vivo

The investigation of the molecular mechanisms mediating the interaction between inflammatory cells and the cells composing the BBB in vivo is hampered by the localization of the brain inside the skull. Most studies addressing the role of CAMs in the development of a CNS inflammatory cellular infiltrate have investigated the effects of monoclonal antibodies directed against CAMs on the evolution of CNS inflammation. Again, most of these studies have been performed in the EAE model. These studies fail to provide definitive evidence that the inhibitory effect of the applied antibody is truly mediated by inhibiting T cell traffic into the CNS, as the antibody might also interfere with the development of the immune response. Interaction of ICAM-1 with LFA-1 on T lymphocytes plays an important role in antigen presentation and recognition, and hence T cell activation and proliferation (reviewed in Carlos and Harlan 1994). Although there is no direct evidence for a role of α4-integrin in antigen dependent T cell stimulation, it has been shown that α4-mediated binding of T lymphocytes to VCAM-1 leads to T cell stimulation (Romanic and Madri 1994).

Antibodies against LFA-1 and Mac-1 have been ineffective in inhibiting clinical EAE and/or leukocyte infiltration of the CNS during EAE (Cannella et al. 1993; Welsh et al. 1993). However, these monoclonal antibodies diminish mononuclear cell recruitment into the CNS during Sindbis virus-induced encephalitis (Irani and Griffin 1996). It has been shown that antibodies directed against ICAM-1 inhibit the development of actively induced EAE (Archelos et al. 1993) but not adoptively transferred EAE (Archelos et al. 1993; Willenborg et al. 1993; Cannella et al. 1993). These results suggests that the inhibitory effect on disease progression which is seen following the administration of anti-ICAM-1 antibody in actively induced EAE is due to the interference with antigen recognition by T cells rather than with T cell trafficking into the CNS. This is further underlined by the findings that treatment of EAE with anti-LFA-1 can increase the severity of the disease (Welsh et al. 1993).

Similarly, although in independent studies it has been observed that repeated injections of the antibody Mel-14, which is directed against L-selectin, have no influence on the development of passively-transferred (t)EAE (Veromaa et al. 1993), this antibody readily inhibits actively-induced (a)EAE (Engelhardt 1997).

Treatment of both aEAE or tEAE with monoclonal antibodies directed against P- or E-selectin does not interfere with the development of the disease, which underlines the lack of expression of both selectins by BBB endothelium in vivo (Engelhardt et al., 1997).

5 The Key Role of α4-Integrin

A key role for α4-integrin mediated T lymphocyte recruitment across the BBB during EAE is supported by the results from several experiments. Autoaggressive T cells have to be freshly activated in order to induce EAE. These freshly activated T line cells express a memory/activated phenotype with high levels of LFA-1 and CD44, low levels of CD45RB and most of them express ICAM-1. There is some evidence that expression of α4-integrin correlates with the disease inducing activity of PLP- and MBP-specific T cell clones (BARON et al. 1993; KUCHROO et al. 1993). KUCHROO and colleagues could demonstrate that most α4-integrin-positive PLP-specific T cell clones were able to induce EAE in SJL/J mice, while α4-negative PLP-specific clones did not mediate EAE. Interestingly, α4-integrin-negative PLP-specific T cells were able to induce EAE in irradiated recipients. It should be noted that the authors also found some α4-integrin-positive T cell clones which did not induce EAE, probably due to an altered cytokine profile. These authors concluded that α4-integrin on the surface of PLP-specific T cells might be important for their migration across the BBB and that dependence of α4-integrin for immigration of T cells into the CNS is abolished by irradiation of the recipients. In another study, BARON et al. (1993) showed that α4-integrin expression on MBP-specific T cells directly correlated with their entry into the brain parenchyme and their ability to transfer EAE into irradiated PL/J mice. The α4-integrin levels on these MBP-specific T cells did not affect antigen responsiveness or production of the Th1 cytokines IL-2, interferon (IFN)-γ, and lymphotoxin, and antibodies directed against α4-integrin did not block antigen recognition in vitro. Thus these authors concluded that α4-integrin on activated effector T cells is crucial for leaving the bloodstream and entering the brain. It is noteworthy that, in contrast to the study by KUCHROO et al., BARON and colleagues saw α4-integrin dependence of T cell recruitment into the CNS in irradiated recipients. Further studies will be needed to define the apparent differences.

Antibodies against α4-integrin have been shown to have therapeutic effects in the adoptive transfer model of EAE in the rat (YEDNOCK et al. 1992) and the mouse (BARON et al. 1993). YEDNOCK et al. showed that intraperitoneal injection of anti-α4-integrin antibody on day 2 after passive transfer of EAE significantly delayed onset of clinical disease and prevented the accumulation of leukocytes in the CNS. As the antibody was given after initiation of disease it was proposed that monoclonal antibody treatment blocked entry of host mononuclear cells recruited to the site of inflammation. Further studies in the actively induced guinea pig EAE model confirmed and extended these results (KESZTHELYI et al. 1996; KENT et al. 1995). Importantly, in these studies treatment with anti-α4 antibody after the onset of disease reversed the clinical symptoms of EAE and resulted in the clearance of leukocytes from the CNS. Since further accumulation of leukocytes could be decreased by anti-α4-integrin antibodies, it appears that individual leukocytes are present in the lesion area for only a limited period of time. The data from these studies are thus consistent with a central role for α4-integrin on antigen-specific T cells and on nonspecific leukocytes for recruitment into neural tissue. This means that trafficking across a leaky BBB during later stages of the disease is still dependent upon α4-integrin interactions.

Little is known about the resolution of lesions in EAE. Present data indicate that leukocytes must either traffic rapidly out of the CNS (reviewed by WELLER et al. 1996) or undergo apoptosis (SCHMIED et al. 1993). Therefore, possible effects of α4-integrin antibodies on T cell clearance inside the CNS have to be considered, although there is no evidence to date indicating that α4-integrin might be involved in induction of T cell apoptosis. As all the antibody inhibition studies have been performed using whole immunoglobulin, the effector functions of the applied antibodies have to be considered when interpreting their effects in vivo (HAMANN et al. 1994). Additionally, anti-α4-integrin antibodies could affect T cell activation (DAMLE and ARUFFO 1991) and enzyme secretion (ROMANIC and MADRI 1994). Similar efficacy of VCAM-1-directed treatment in EAE (BARON et al. 1993; Lobb, Burkly and Ruddle, unpublished), however, argues for a mechanism that involves blockade of adhesion dependent functions, but further examination of these issues will be of value.

The striking ability of antibodies against α4-integrin to inhibit CNS inflammation in EAE is in contrast to the more limited ability of these reagents to prevent mononuclear cell infiltration in other organs. In models of rheumatoid arthritis, cardiac allograft rejection, lung inflammation and skin inflammation antibodies to α4-integrin either failed to inhibit or only partially inhibited mononuclear cell traffic (reviewed by LOBB and HEMLER 1994). This stresses the specificity for α4-integrin-mediated cell recruitment into the CNS, which so far only seems to be similar in the gut, where mononuclear cell inflitration has been demonstrated to be dependent on α4-integrin (ISSEKUTZ 1991). The relevant integrin in the gut however is likely to be α4β7 (HAMANN et al. 1994) rather than α4β1. It should be stressed at this point that, besides demonstrating the lack of α4β7-integrin expression on inflammatory cells in the CNS (ENGELHARDT et al. 1995; ENGELHARDT et al., submitted), the respective functions of α4β1 vs α4β7 in trafficking into the CNS have not been completely elucidated. This is especially important as α4β7 can also mediate binding to VCAM-1.

The importance of α4-integrin in leukocyte recruitment across the BBB becomes even more compelling, as α4-integrins have been demonstrated to have a unique dual adhesive function. Besides their activation dependent binding to their respective ligands VCAM-1 and MAdCAM-1, α4-integrins have been shown to interact with both of these ligands under flow in vitro (BERLIN et al. 1995; ALON et al. 1995). Both studies provide evidence that α4-mediated rolling on VCAM-1 or MAdCAM-1 did not require integrin activation although activation of α4-integrin by Mn^{2+} increased the number of rolling cells but also favored firm adhesion. Whereas α4β7-mediated rolling on MAdCAM-1 could be observed under flow rates comparable to those in which selectin-mediated interactions take place, α4β1-mediated rolling on VCAM-1 was observed under conditions with lower shear forces (ALON et al. 1995). In vivo there is evidence that α4-integrin can mediate rolling and adhesion, but not tethering under physiological shear (4–16 dynes/cm^2) in a model of chronic vasculitis of the mesenterium (JOHNSTON et al. 1996). The possibility that α4-integrin mediates multiple leukocyte interactions such as rolling and adhesion at the BBB remains to be investigated.

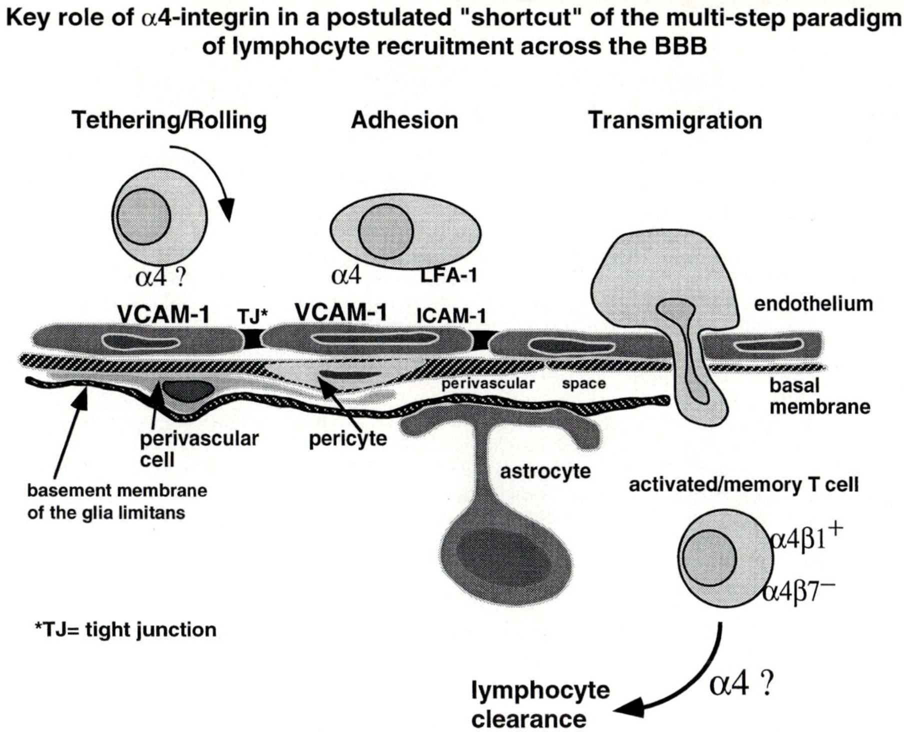

Fig. 1. Key role of α4-integrin in a postulated shortcut of the multistep paradigm of lymphocyte recruitment across the blood-brain barrier

6 Conclusion

Taken together: (1) only ICAM-1 and VCAM-1 but not E- and P-selectin are upregulated on BBB endothelium during inflammatory conditions of the CNS. (2) Anti-α4-integrin antibodies interfere with the development of inflammatory infiltrates in the CNS in vivo and inhibit the development of clinical EAE. (3) α4-integrin has a unique dual function in binding to its ligands VCAM-1 and MAdCAM-1 under physiological flow as well as under static conditions. These data provoke the speculation that recruitment of mononuclear cells across the BBB could be controlled by a shortcut of the multistep paradigm (Fig. 1), limiting access to the CNS to those cells that are capable of α4-integrin-mediated tethering and rolling followed by firm adhesion to VCAM-1 on the luminal surface of the BBB endothelium.

Acknowledgements. I owe thanks to Sara Michie with Badger for discussion of this present review.

References

Alon R, Kassner PD, Carr MW, Finger EB, Hemler ME, Springer TA (1995) The integrin VLA-4 supports tethering and rolling in flow on VCAM-1. J Cell Biol 128:1243–1253

Andersson PB, Perry VH, Gordon S (1992a) Intracerebral injection of proinflammatory cytokines or leukocyte chemotaxins induces minimal myelomonocytic cell recruitment to the parenchyma of the central nervous system. J Exp Med 176:255–259

Andersson PB, Perry VH, Gordon S (1992b) The acute inflammatory response to lipopolysaccharide in CNS parenchyma differs from that in other body tissues. Neuroscience 48:169–186

Archelos JJ, Jung S, Maurer M, Schmied M, Lassmann H, Tamatani T, Miyasaka M, Toyka KV, Hartung HP (1993) Inhibition of experimental autoimmune encephalomyelitis by an antibody to the intercellular adhesion molecule ICAM-1. Ann Neurol 34:145–154

Bargatze RF, Jutila MA, Butcher EC (1995) Distinct roles of L-selectin and integrins $\alpha 4\beta 7$ and LFA-1 in lymphocyte homing to peyer's patch-HEV in situ: the multistep model confirmed and refined. Immunity 3:99–108

Barkalow F, Goodman M, Gerritsen M, Mayadas TN (1996) Brain endothelium lack one of two pathways of P-selectin – mediated neutrophil adhesion. Blood 88:4585–4593

Baron JL, Madri JA, Ruddle NH, Hashim G, Janeway CA Jr (1993) Surface expression of alpha 4 integrin by CD4 T cells is required for their entry into brain parenchyma. J Exp Med 177:57–68

Barten DM, Clark RB, Ruddle NH (1995) Presence of T cells with activated and memory phenotypes in inflammatory spinal cord lesions. J Immunol 155:5409–5418

Berlin C, Bargatze RF, Campbell JJ, von Andrian UH, Szabo MC, Hasslen SR, Nelson RD, Berg EL, Erlandsen SL, Butcher EC (1995) $\alpha 4$ integrins mediate lymphocyte attachment and rolling under physiologic flow. Cell 80:413–422

Breier G, Breviaro F, Caveda L, Berthier R, Schnürch H, Gotsch U, Vetweber D, Risau W, Dejana E (1995) Molecular cloning and expression of murine VE-cadherin in early developing cardiovascular system. Blood 87:630–641

Butcher EC (1991) Leukocyte-endothelial cell recognition: three (or more) steps to specificity and diversity. Cell 67:1033–1036

Cannella B, Cross AH, Raine CS (1990) Upregulation and coexpression of adhesion molecules correlate with relapsing autoimmune demyelination in the central nervous system. J Exp Med 172:1521–1524

Cannella B, Cross AH, Raine CS (1991a) Adhesion-related molecules in the central nervous system. Upregulation correlates with inflammatory cell influx during relapsing experimental autoimmune encephalomyelitis. Lab Invest 65:23–31

Cannella B, Cross AH, Raine CS (1991b) Relapsing autoimmune demyelination: a role for vascular addressins. J Neuroimmunol 35:295–300

Cannella B, Cross AH, Raine CS (1993) Anti-adhesion molecule therapy in experimental autoimmune encephalomyelitis. J Neuroimmunol 46:43–55

Carlos TM, Harlan JM (1994) Leukocyte endothelial adhesion molecules. Blood 84:2068–2101

Chou C-H, Shapira R, Fritz RB (1983) Encephalitogenic activity of small form of myelin basic protein in the SJL/J mouse. J Immunol 130:2183

Damle NK, Aruffo A (1991) Vascular cell adhesion molecule 1 induces T cell antigen receptor dependent activaiotn of CD4+ T lymphocytes. Proc Natl Acad Sci USA 88:6403–6407

Deckert-Schlüter M, Schlüter D, Hof H, Wiestler OD, Lassmann H (1994) Differential expression of ICAM-1, VCAM-1 and their ligands LFA-1, Mac-1, CD43, VLA-4, and MHC class II antigens in murine Toxoplasma encephalitis: a light microscopic and ultrastructural immunohistochemical study. J Neuropathol Exp Neurol 53:457–468

Dopp JM, Breneman SM, Olshcowka JA (1994) Expression of ICAM-1, VCAM-1, L-selectin, and leukosialin in the mouse central nervous system during the induction and remission stages of experimental allergic encephalomyelitis. J Neuroimmunol 54:129–144

Engelhardt B, Vestweber D, Hallmann R and Schulz M (1997) -E- and P-selectin are not involved in the recruitment of inflammatory cells across the blood-brain barrier in experimental autoimmune encephalomyelitis. Blood, 90, (in press)

Engelhardt B (1997) The blood-brain barrier. In: Russell WC (ed) Molecular biology of multiple slecrosis. Wiley, Chichester

Engelhardt B, Risau W (1995) The development of the blood-brain barrier. In: Greenwood J, Begley D, Segal M (eds) New concepts of a blood-brain barrier. Plenum, London

Engelhardt B, Conley FK, Butcher EC (1994) Cell adhesion molecules on vessels during inflammation in the mouse central nervous system. J Neuroimmunol 51:199–208

Engelhardt B, Conley FC, Kilshaw PJ, Butcher EC (1995) Lymphocytes infiltrating the CNS during inflammation display a distinctive phenotype and bind to VCAM-1 but not to MAdCAM-1. Int Immunol 7:481–491

Griffin DE, Levine B, Tylor WR, Irani DN (1992) The immune response in viral encephalitis. Semin Immunol 4:111–119

Hamann A, Andrew DP, Jablonski-Westrich D, Holzmann B, Butcher EC (1994) Role of alpha 4-integrins in lymphocyte homing to mucosal tissues in vivo. J Immunol 152:3282–3293

Hickey WF, Hsu BL, Kimura H (1991) T-lymphocyte entry into the central nervous system. J Neurosci Res 28:254–260

Hurwitz AA, Lyman WD, Guida MP, Calderon TM, Berman JW, Jensen MA, Arnason BG, Toscas A, Noronha A (1992) Tumor necrosis factor alpha induces adhesion molecule expression on human fetal astrocytes: preferential increase of IL-2R+ CD4+ T cells and CD45RB- CD4+ T cells in the central nervous system in experimental allergic encephalomyelitis. J Exp Med 176:1631–1636

Irani DN, Griffin DE (1996) Regulation of lymphocyte homing into the brain during viral encephalitis at various stages of infection. J Immunol 156:3850–3857

Issekutz TB (1991) Inhibition of in vivo lymphocyte migration to inflammation and homing to lymphoid tissues by the TA-2 monoclonal antibody. A likely role for VLA-4 in vivo. J Immunol 147:4178–4184

Johnston B, Issekutz TB, Kubes P (1996) Tha α4-integrin supports leukocyte rolling and adhesion in chronically inflamed postcapillary venules in vivo. J Exp Med 183:1995–2006

Jung S, Toyka K, Hartung H-P (1995) Suppression of experimental autoimmune encephalomyelitis in Lewis rats by antibodies against CD2. Eur J Immunol 25:1391–1398

Kent SJ, Karlik SJ, Cannon C, Hines DK, Yednock TA, Fritz LC, Horner HC (1995) A monoclonal antibody to alpha 4 integrin suppresses and reverses active experimental allergic encephalomyelitis. J Neuroimmunol 58:1–10

Keszthelyi E, Karlik S, Hyduk S, Rice GPA, Gordon G, Yednock T, Horner H (1996) Evidence for a prolonged role of a4 integrin throughout active experimental allergic encephalomyelitis. Neurol 47:1053–1059

Kuchroo VK, Martin CA, Greer JM, Ju S-T, Sobel RA, Dorf ME (1993) Cytokines and adhesion molecules contribute to the ability of myelin proteolipid protein-specific T cell clones to mediate experimental allergic encephalomyelitis. J Immunol 151:4371–4382

Lassmann H, Rossler K, Zimprich F, Vass K (1991) Expression of adhesion molecules and histocompatibility antigens at the blood-brain barrier. Brain Pathol 1:115–123

Lees MB, Kuchroo VK, Sobel RA (1989) Myelin proteolipid protein: its roal in experimental allergic encephalomyelitis. Int Pediatr 6:84

Lenter M, Uhlig H, Hamann A, Jeno P, Imhof B, Vestweber D (1993) A monoclonal antibody against an activation epitope on mouse integrin chain beta 1 blocks adhesion of lymphocytes to the endothelial integrin alpha 6 beta 1. Proc Natl Acad Sci USA 90:9051–9055

Liebert UG, ter Meulen V (1993) Synergistic interaction between measles virus infection and myelin basic protein peptide-specific T cells in the induction of experimental allergic encephalomyelitis in Lewis rats. J Neuroimmunol 46:217–223

Lobb RR, Hemler ME (1994) The pathophysiological role of a4 integrins in vivo. J Clin Invest 94: 1722–1728

Mackay CR (1991) T-cell memory: the connection between function, phenotype and migration pathways. Immunol Today 12:189–192

Martin R, McFarland HF (1995) Immunological aspects of experimental allergic encephalomyelitis and multiple sclerosis. Crit Rev Clin Laor Sci 32:121–182

Musette P, Burcau J-F, Gachelin G, Kourilsky P, Brahic M (1995) T lymphocyte repertoire in Theiler's virus encephalomyelitis: the nonspecific inflitraiton of the central nervous system of infected SJL/J mice is asociated with a selective local T cell expansion. Eur J Immunol 25:1589–1593

Nagy Z, Peters H, Hüttner I (1984) Fracture faces of cell junctions in cerebral endothelium during normal and hyperosmotic conditions. Lab Invest 50:313–322

Nico B, Cantino D, Bertossi M, Ribatti D, Sassoe M, Roncali L (1992) Tight endothelial junctions in the developing microvasculature – a thin-section and freeze-fracture study in the chick-embryo optic tectum. J Submicr Cytol 24:85–95

O'Neill JK, Butter C, Baker D, Gschmeissner SE, Kraal G, Butcher EC, Turk JL (1991) Expression of vascular addressins and ICAM-1 by endothelial cells in the spinal cord during chronic relapsing experimental allergic encephalomyelitis in the Biozzi AB/H mouse. Immunology 72:520–525

Picker LJ, Treer JR, Ferguson DB, Collins PA, Bergstresser PR, Terstappen LW (1993) Control of lymphocyte recirculation in man. II. Differential regulation of the cutaneous lymphocyte-associated antigen, a tissue-selective homing receptor for skin-homing T cells. J Immunol 150:1122–1136

Raine CS, Lee SC, Scheinberg LC, Duijvestin AM, Cross AH (1990) Adhesion molecules on endothelial cells in the central nervous system: an emerging area in the neuroimmunology of multiple sclerosis. Clin Immunol Immunopathol 57:173–187

Renno T, Zeine R, Girard JM, Gillani S, Dodelet V, Owens T (1994) Selective enrichment of Th1 CD45RBlow CD4+ T cells in autoimmune infiltrates in experimental allergic encephalomyelitis. Int Immunol 6:347–354

Richt JA, Stitz L, Wekerle H, Rott R (1989) Borna disease, a progressive menigoencephalomyelitis as a model for CD4+ T cell-meidated immunopathology in the brain. J Exp Med 170:1045–1050

Romanic AM, Madri JA (1994) The induction of 72-kd gelatinase in T-cells upon adhesion to endothelial-cells is VCAM-1 dependent. J Cell Biol 125(N5):1165–1178

Rössler K, Neuchrist C, Kitz K, Scheiner O, Kraft D, Lassmann H (1992) Expression of leucocyte adhesion molecules at the human blood-brain barrier (BBB). J Neurosci Res 31:365–374

Sasseville VG, Newman WA, Lackner AA, Smith MO, Lausen NC, Beall D, Ringler DJ (1992) Elevated vascular cell adhesion molecule-1 in AIDS encephalitis induced by simian immunodeficiency virus. Am J Pathol 141:1021–1030

Sasseville VG, Newman W, Brodie SJ, Hesterberg P, Pauley D, Ringler DJ (1994) Monocyte adhesion to endothelium in simian immunodeficiency virus-induced aids encephalitis is mediated by vascular cell-adhesion molecule-1/alpha-4-beta-1 integrin interactions. Am J Pathol 144(N1):27–40

Schmied M, Breitschopf H, Gold R, Zischler H, Rothe G, Wekerle H, Lassmann H (1993) Apoptosis of T lymphocytes in experimental autoimmune encephalomyelitis. Evidence for programmed cell death as a mechanism to control inflammation in the brain. Am J Pathol 143:446–452

Springer TA (1994) Traffic signals for lymphocyte recirculation and leukocyte emigration: the multistep paradigm. Cell 76:301–314

Stamper HBJ, Woodruff JJ (1976) Lymphocyte homing into lymph nodes: in vitro demonstration of the selective affinity of recirculating lymphocytes for high endothelial venules. J Exp Med 144:828

Steffen BJ, Butcher EC, Engelhardt B (1994) Evidence for involvement of ICAM-1 and VCAM-1 in lymphocyte interaction with endothelium in experimental autoimmune encephalomyelitis in the central nervous system in the SJL/J mouse. Am J Pathol 145(N1):189–201

Tsunoda I, Fujinami RS (1996) Two models for multiple sclerosis: experimental allergic encephalomyelitis and Theiler's murine encephalomyelitis virus. J Neuropathol Exp Neurol 55:673–686

Veromaa T, Weissman IL, Steinman L, Bocke S (1993) A critical role for integrin a4 and CD44 in the induction of experimental autoimmune encephalomyelitis (EAE) in mice. J Cell Biochem 17A:356

Washington R, Burton J, Todd RFR, Newman W, Dragovic L, Dore-Duffy P (1994) Expression of immunologically relevant endothelial cell activation antigens on isolated central nervous system microvessels from patients with multiple sclerosis. Ann Neurol 35:89–97

Wekerle H, Linington C, Lassmann H, Meyermann R (1986) Cellular immune reactivity within the CNS. TINS 9:271–277

Weller RO, Engelhardt B, Phillips MJ (1996) Lymphocyte targeting of the central nervous system: a review of afferent and efferent CNS-immune pathways. Brain Pathol 6:275–288

Welsh CT, Rose JW, Hill KE, Townsend JJ (1993) Augmentation of adoptively transferred experimental allergic encephalomyelitis by administration of a monoclonal antibody specific for LFA-1a. J Neuroimmunol 43:161–168

Willenborg DO, Simmons RD, Tamatani T, Miyasaka M (1993) ICAM-1-dependent pathway is not critically involved in the inflammatory process of autoimmune encephalomyelitis or in cytokine-induced inflammation of the central nervous system. J Neuroimmunol 45:147–154

Wolburg H, Neuhaus J, Kniesel U, Krauss B, Schmid E-M, Öcalan M, Farrell C, Risau W (1994) Modulation of tight junction structure in blood-brain barrier endothelial cells. J Cell Sci 107: 1347–1357

Yednock TA, Cannon C, Fritz LC, Sanchez Madrid F, Steinman L, Karin N (1992) Prevention of experimental autoimmune encephalomyelitis by antibodies against alpha 4 beta 1 integrin. Nature 356:63–66

Zeine R, Owens T (1992) Direct demonstration of the infiltration of murine central nervous system by Pgp-1/CD44high CD45RBlow CD4+ T cells that induce experimental allergic encephalomyelitis. J Neuroimmunol 40:57–70

The Roles of α4-Integrins in the Development of Insulin-Dependent Diabetes Mellitus

S.A. Michie[1,2], H.-K. Sytwu[3], J.O. McDevitt,[3,4] and X.-D. Yang[5]

1 Introduction . . . 66

2 Leukocyte α4-Integrins and Their Endothelial Ligands . . . 67

3 Leukocyte α4-Integrins in Inflammation . . . 68

4 α4-Integrins in Insulin-Dependent Diabetes Mellitus: In Vitro Studies from the NOD Mouse . . . 69

5 α4-Integrins in Insulin-Dependent Diabetes Mellitus: In Vivo Studies from the NOD Mouse . . . 71
5.1 Direct Inhibition of Lymphocyte Migration Into Islets . . . 71
5.2 Prevention of Insulitis and Diabetes . . . 72

6 Other Lymphocyte/Endothelial Adhesion Systems in NOD Diabetes . . . 73
6.1 The Peripheral Node Adhesion System . . . 73
6.2 LFA-1 and ICAM-1 . . . 74

7 The Roles of Cytokines in Pancreatic Endothelial Adhesion Molecule Expression and Inflammation . . . 74

8 α4-Integrins and Their Ligands in Human Insulin-Dependent Diabetes Mellitus . . . 75

9 α4-Integrins in NOD Inflamed Salivary Gland and Lacrimal Gland: Comparison to NOD Inflamed Pancreas . . . 76

10 α4-Integrins in Experimental Allergic Encephalomyelitis: Comparison to NOD Diabetes . . . 76

11 Caveats . . . 77

12 Summary . . . 79

References . . . 79

[1]Department of Veterans Affairs, Palo Alto Health Care System, Center for Molecular Biology in Medicine, Palo Alto, CA, 94304, USA
[2]Department of Pathology, Stanford University School of Medicine, Stanford, CA, 94305–5324, USA
[3]Department of Microbiology and Immunology, Stanford University School of Medicine, Stanford, CA, 94305–5402, USA
[4]Department of Medicine, Stanford University School of Medicine, Stanford, CA, 94305–5111, USA
[5]Division of Preclinical Research, Abgenix, Inc., Fremont, CA, 94555, USA

1 Introduction

Most lymphocytes recirculate throughout the body, migrating from blood through organized lymphoid tissues such as lymph nodes (LN) and Peyer's patches (PP), then to lymph and back to blood (Gowans and Knight 1964). Smaller numbers of lymphocytes migrate from blood to extranodal tissues such as pancreas and then through lymphatic vessels to LN (Mackay et al. 1990). An important feature of this migration is the ability of lymphocytes to recognize and adhere to the surface of blood vessel endothelial cells before migrating through the vessel wall into surrounding tissue (Carlos and Harlan 1994; Imhof and Dunon 1995; Butcher and Picker 1996).

Adhesion interactions of vascular endothelium with lymphocytes under flow or shear consist of at least four steps: (1) an initial transient sticking or rolling; (2) if the lymphocytes encounter appropriate activating or chemotactic factors in the local environment, rolling may be followed by a lymphocyte activation step that then leads to; (3) strong adhesion or sticking that may be followed by; (4) lymphocyte diapedesis into tissue (Butcher 1991; Shimuzu et al. 1992; Springer 1994; Bargatze et al. 1995). Specific lymphocyte and endothelial adhesion molecules (AM) are involved in each step of this "adhesion cascade" (reviewed in Carlos and Harlan 1994; Imhof and Dunon 1995; Butcher and Picker 1996). This allows lymphocyte migration to be controlled at several different steps, leading to a combinatorial increase in specificity and sensitivity.

Some lymphocyte and endothelial AM, such as lymphocyte function associated antigen-1 (LFA-1), appear to play a role in lymphocyte migration to a wide variety of normal and inflamed tissues (Hamann et al. 1988; Carlos and Harlan 1994; Bargatze et al. 1995). Other AM are involved in migration in a more selective fashion. For example, in the mouse, well-defined tissue-selective lymphocyte/endothelial adhesion systems exist for peripheral LN (PLN) and mucosal lymphoid tissues such as PP. Lymphocyte L-selectin and its endothelial carbohydrate ligands, presented on different glycoprotein backbones collectively known as the peripheral node addressin (PNAd), play a key role in lymphocyte migration to PLN (Gallatin et al. 1983; Streeter et al. 1988a; Baumhueter et al. 1992; Lasky et al. 1992). Although L-selectin and PNAd are also involved in lymphocyte migration to PP, specific homing to this site appears to be dominated by lymphocyte $\alpha 4\beta 7$ binding to endothelial mucosal addressin cell adhesion molecule-1 (MAdCAM-1) (Streeter et al. 1988b; Holzmann et al. 1989; Hu et al. 1992; Berlin et al. 1993; Briskin et al. 1993; Hamann et al. 1994; Bargatze et al. 1995). Other AM, such as vascular cell adhesion molecule-1 (VCAM-1), may mediate lymphocyte migration to a wide variety of inflammatory sites (reviewed in Postigo et al. 1993; Lobb and Hemler 1994).

Lymphocyte migration is important in immune surveillance in that it allows lymphocytes to encounter their specific antigen in almost any site of antigen entry or sequestration. Such encounters often lead to localized inflammation characterized by recruitment of effector leukocytes that mediate destruction of the antigen.

Ideally, this is accomplished with little damage to normal host tissues. However, lymphocyte migration can lead to persistent chronic inflammation that damages or destroys host tissues. This inflammation can be induced by exogenous antigens such as bacteria, or by an autoimmune response to self antigens. Inflammation in autoimmune diseases is often limited to one, or a small number, of tissues. Insulin-dependent diabetes mellitus (IDDM) is one clinically important example of such a tissue-specific, autoimmune disease (Castano and Eisenbarth 1990; Rossini et al. 1993; Bach 1994). In IDDM, there is an autoimmune response to antigens in the insulin-producing β-cells of the pancreatic islets of Langerhans. This leads to lymphocytic infiltration of the islets (insulitis) and subsequent destruction of the β-cells.

Investigation of the importance of lymphocyte migration in autoimmune diseases such as IDDM has involved a series of experiments covering different time points during the disease process. These experiments have used animal models to delineate the lymphocyte/endothelial adhesion pathways involved in the diabetogenic process. The nonobese diabetic (NOD) mouse offers an ideal model for these studies because: (1) NOD mice spontaneously develop IDDM that closely resembles human IDDM (Castano and Eisenbarth 1990); (2) NOD mice develop lymphocytic infiltrates in several organs including pancreas, lacrimal gland and salivary gland, allowing comparison of different adhesion pathways in the same animal (Asamoto et al. 1984); and (3) monoclonal antibodies (mAbs) are available against a wide variety of murine lymphocyte and endothelial AM. Understanding the mechanisms that regulate lymphocyte migration to the pancreatic islets has led to further understanding of the pathogenesis of IDDM and provides a rational basis for the development of AM-based therapies for IDDM. In this chapter, we summarize the recent advances regarding the functions of leukocyte α4-integrins, their endothelial ligands, and other AM in the development of IDDM in the NOD mouse. These results are compared to the functions of lymphocyte and endothelial AM in the development of experimental allergic encephalomyelitis (EAE), an autoimmune disease that results in inflammation of the brain and spinal cord.

2 Leukocyte α4-Integrins and Their Endothelial Ligands

α4-Integrins are expressed by most lymphocytes, NK cells, monocytes, eosinophils and basophils. The α4 chain can pair with either of two β chains to form α4β1 or α4β7 integrin (reviewed in Hynes 1992). Both integrins can exhibit a variety of activation states, based in part upon external stimuli and the cell's differentiation (Hynes 1992). Lymphocyte α4β1 integrin (VLA-4, CD49d/CD29) has at least two major ligands: endothelial vascular cell AM-1 (VCAM-1; CD106) and the extracellular matrix protein fibronectin (Wayner et al. 1989; Elices et al. 1990; Chan et al. 1992). VCAM-1, a member of the immunoglobulin (Ig) superfamily, is constitutively expressed at low levels by some endothelial cells in a variety of tissues including exocrine pancreas, brain, aorta, lung and kidney (RICE et al. 1991; FRIES et al. 1993; HANNINEN et al. 1993; FAVEEUW et al. 1994; STEFFEN et al. 1994). It is expressed at higher levels by endothelium lining the bone marrow sinusoids (JACOBSEN et al. 1996). VCAM-1 is also found on nonendothelial cells such as bone marrow stromal cells, dendritic cells in lymphoid tissues, and some macrophages (MIYAKE et al. 1991; RICE et al. 1991).

Like α4β1, α4β7 binds to VCAM-1 and fibronectin (RUEGG et al. 1992). However, in vitro studies suggest α4β1 predominates over α4β7 in leukocyte adhesion to VCAM-1 or fibronectin (CHAN et al. 1992). In contrast, endothelial MAdCAM-1 is a ligand for α4β7 but not α4β1 (BERLIN et al. 1993). MAdCAM-1, a glycoprotein with several Ig-like domains and a single mucin domain, is constitutively expressed by vessels in PP, mesenteric LN (MLN), intestinal lamina propria, lactating mammary gland, and exocrine pancreas (STREETER et al. 1988b; BRISKIN et al. 1993; BRISKIN et al., submitted; HANNINEN et al. 1993; FAVEEUW et al. 1994). α4β7 and MAdCAM-1 play key roles in lymphocyte migration to mucosal tissues such as PP and MLN but not to PLN (STREETER et al. 1988b; HOLZMANN et al. 1989; HAMANN et al. 1994; BARGATZE et al. 1995). Thus α4β7 and MAdCAM-1 are key components of the mucosal lymphocyte/endothelial adhesion system. MAdCAM-1 is also expressed on marginal zone sinus-lining cells in spleen and follicular dendritic cells in LN and PP (KRAAL et al. 1995; SZABO et al., 1997). Thus MAdCAM-1 may be involved in immune system interactions other than lymphocyte adherence to vascular endothelium.

3 Leukocyte α4-Integrins in Inflammation

A variety of in vitro data initially suggested that leukocyte α4-integrins and endothelial VCAM-1 and MAdCAM-1 are involved in leukocyte migration to sites of inflammation. Several inflammatory mediators, including lipopolysaccharide (LPS), tumor necrosis factor-α (TNF-α), interleukin (IL)-1, and IL-4, up-regulate VCAM-1 expression on cultured endothelial cells (reviewed in CARLOS and HARLAN 1994). This up-regulation results in increased binding of mononuclear cells to the endothelium. Expression of endothelial MAdCAM-1 can be induced by inflammatory mediators (SIKORSKI et al. 1993). In addition, activation of leukocytes by a variety of stimuli can enhance the affinity of α4-integrins for their ligands (HYNES 1992).

Immunohistochemical staining of tissues from rodents and nonhuman and human primates shows VCAM-1 up-regulation on vessels in many different inflamed tissues including skin, lung, synovium, brain, pancreas, gastrointestinal tract, salivary gland, lacrimal gland and kidney (KOCH et al. 1991; RICE et al. 1991; FRIES et al. 1993; BARON et al. 1994; STEFFEN et al. 1994; YANG et al. 1994a; TSUKAMOTO et al. 1995; HUNGER et al. 1996). In contrast, strong expression of endothelial MAdCAM-1 in inflamed tissues from immunocompetent animals has only been reported in the gastrointestinal tract, pancreas and thymus (Hanninen et al. 1993; Faveeuw et al. 1994; Michie et al. 1995; Briskin et al., submitted). This suggests that MAdCAM-1 could be a ligand for lymphocyte migration to a select group of inflamed tissues.

Several rodent and a few nonhuman primate models have been used to examine the roles of α4-integrins and their ligands in leukocyte migration into extranodal inflammatory sites. These models include allograft rejection, graft vs host disease, contact hypersensitivity, pulmonary allergic reactions, arthritis, colitis and thymic hyperplasia. Most of these studies have used mAbs against the α4 chain or VCAM-1 to demonstrate that these AM play important roles in lymphocyte migration in vivo. Several excellent reviews are available that give more detailed information about these studies (Carlos and Harlan 1994; Postigo et al. 1993; Lobb and Hemler 1994). In addition, the few in vivo studies examining the functions of β7 or MAdCAM-1 in chronic inflammation support roles for these AM in lymphocyte migration to sites of inflammation (Michie et al. 1995; Hesterberg et al. 1996; Picarella et al. 1997; Yang et al., in press).

4 α4-Integrins in Insulin-Dependent Diabetes Mellitus: In Vitro Studies from the NOD Mouse

Immunohistochemical studies have been used to define endothelial AM expression in NOD pancreas. MAdCAM-1 is found on many vessels in exocrine pancreas of NOD mice from birth through old age (Hanninen et al. 1993; Faveeuw et al. 1994). Mononuclear inflammatory cells are first detected next to islets at about 3 weeks of age. Concurrently, vessels with the morphology of high endothelial venules (HEV), which are the vessels involved in most lymphocyte migration from blood into LN and PP, develop next to the islets (Michie, unpublished data). From 3 weeks of age, increasing numbers of mononuclear cells including T and B lymphocytes accumulate around the islets (peri-insulitis) and gradually invade into the islets (insulitis). MAdCAM-1 seems to be the predominant addressin expressed on endothelium next to the islets during the initial stages of insulitis (3–6 weeks) (Table 1) (Hanninen et al. 1993; Faveeuw et al. 1994). In contrast, very little PNAd expression (see Sect. 6.1) is seen until there is significant insulitis (approximately 8 weeks of age). As insulitis progresses, there is an increase in the number of vessels expressing MAdCAM-1 and PNAd (Yang et al. 1994a) (Table 1). Both of these addressins are expressed mainly by peri-islet HEV.

VCAM-1 is expressed by vessels in exocrine and endocrine pancreas of NOD mice (Hanninen et al. 1993; Lo et al. 1993; Faveeuw et al. 1994; Baron et al. 1994; Yang et al. 1994a; Tsukamoto et al. 1995). In very young mice without insulitis, a few VCAM-1 expressing vessels are found next to the islets whereas many are found in the exocrine pancreas (Hanninen et al. 1993; Faveeuw et al. 1994). In older mice with insulitis, VCAM-1 expression is up-regulated on vessels in the areas

Table 1. AM expression on lymphocytes and vascular endothelium in islets of NOD mice[a]

Age (weeks)	Lymphocyte AM[b]		Endothelial AM[c]		
	α4β7	L-selectin	MAdCAM-1	VCAM-1	PNAd
3	+ + +	+	±	±	0
4	+ + +	+	±	±	±
5	+ + +	+	+	±	±
6	+ + +	+	+	±	±
8	+ + +	+ +	+ +	±	+
12	+ + +	+ +	+ + +	+	+ +
14–17	+ + +	+ +	+ + +	+	+ +
18–30	+ + +	+ +	+ + +	+ +	+ +

[a]AM expression was evaluated on immunoperoxidase-stained sections of inflamed pancreata from female NOD mice (HANNINEN et al. 1993; YANG et al. 1994a; MICHIE, unpublished data).
[b]Expression of α4β7 (mAb DATK-32, R1-2 and Fib504) and L-selectin (mAb MEL-14): +, ≤33%; + +, 34%–66%, + + +, ≥67% of infiltrating cells in areas of insulitis express the AM.
[c]Expression of MAdCAM-1 (mAb MECA-367), VCAM-1 (mAb M/K-2.7) and PNAd (mAb MECA-79): 0, 0%; ±, 1%–5%; +, 6%–15%; + +, 16%–60%; + + +, ≥61% of islets show endothelial expression of the AM.

of inflammation (Table 1) (YANG et al. 1994a; TSUKAMOTO et al. 1995). VCAM-1 is also expressed on some dendritic cells in inflamed islets.

The vessels in exocrine pancreas of nondiabetes prone mice such as BALB/c and SJL express MAdCAM-1 and VCAM-1 in a pattern similar to that seen in exocrine pancreas of NOD mice (HANNINEN et al. 1993; WOGENSEN et al. 1993; LEE and SARVETNICK 1994; FAVEEUW et al. 1994). Specifically, both AM are seen in exocrine pancreas on a small number of vessels with flat endothelium. The nondiabetes prone mice do not develop pancreatic inflammation, peri-islet HEV, or peri-islet vascular addressin expression.

Immunohistochemical stains of inflamed pancreata from our NOD colony show that most lymphocytes in islets express α4 (mAb R1-2), β7 (mAb Fib504), and α4β7 (mAb DATK-32, which reacts specifically with the α4β7 heterodimer) (Table 1) (HANNINEN et al. 1993; YANG et al. 1994a). The expression of α4β7 by lymphocytes in islets correlates with high expression of MAdCAM-1 by endothelium. Moreover, we have used in vitro assays of lymphocyte/endothelial binding to show that the MAdCAM-1 expressing peri-islet HEV avidly bind lymphocytes by α4β7- and MAdCAM-1-mediated mechanisms (STAMPER and WOODRUFF 1976; HANNINEN et al. 1993; YANG et al. 1993). These data suggest a prominent role for the mucosal lymphocyte/endothelial adhesion pathway in the development of insulitis.

Two groups have used flow cytometry to examine α4β7 expression on islet-infiltrating lymphocytes (FAVEEUW et al. 1994; HANNINEN et al. 1996). Besides pairing with α4, the β7 chain can pair with αE to form the αEβ7 integrin. αEβ7 is found on most gut intraepithelial lymphocytes but on fewer than 5% of NOD islet-infiltrating lymphocytes (KILSHAW and MURANT 1991; HANNINEN et al. 1996; YANG et al., in press). Thus, although neither group used a mAb that reacts specifically with the α4β7 heterodimer, their anti-β7 staining should give an accurate picture of α4β7 expression in inflamed islets. HANNINEN and colleagues (1996) demonstrate that most islet infiltrating lymphocytes in 8–12 week old NOD mice are $\beta7^{hi}$ while such cells from diabetic mice are $\beta7^{neg/lo}$ (mAbs Fib504, Fib21 and Fib30). In contrast, FAVEEUW found that most inflammatory cells in NOD islets are $\beta7^{neg/lo}$ (mAb M301) (FAVEEUW et al. 1994). This discordance in results may be due to differences in technical procedures such as isolation methods or staining techniques, differences in AM expression between colonies, or differences in ages of the mice examined.

Most islet infiltrating T cells express low levels of L-selectin and high levels of LFA-1 and CD44, a phenotype for activated/memory T cells (HANNINEN et al. 1993, 1996; FAVEEUW et al. 1994; GOLDRATH et al. 1995). β1 expression by these cells has not been examined.

5 α4-Integrins in Insulin-Dependent Diabetes Mellitus: In Vivo Studies from the NOD Mouse

5.1 Direct Inhibition of Lymphocyte Migration Into Islets

In vivo studies using mAbs that recognize lymphocyte α4 or its endothelial ligands have been used to determine which adhesion pathway(s) has a prominent role during the diabetogenic process (Table 2). Only one study has directly examined the ability of anti-AM mAbs to inhibit the migration of lymphocytes from blood into islets of unirradiated prediabetic NOD mice. FAVEEUW and colleagues (1995)

Table 2. Anti-α4 integrin, VCAM-1 and MAdCAM-1 based immunotherapy of diabetes in NOD mice

Target	Inhibition of diabetes		Reference
	Spontaneous	Adoptive transfer	
α4-integrin (mAb R1-2)	+ + +	+ + to + + +	YANG et al. 1993; BARON et al. 1994; BURKLY et al. 1994; YANG et al. 1994a; FABIEN et al. 1996
α4-integrin (mAb P/S2)	+ + +	+ + to + + +	BURKLY et al. 1994; TSUKAMOTO et al. 1995; MICHIE, unpublished data
α4β7-integrin	+ +	ND	MICHIE, unpublished data
β7-integrin	+ + +	ND	YANG et al., in press
β1-integrin	ND	ND	None
MAdCAM-1	+ + +	ND	YANG et al., in press
VCAM-1	+/–	+/– to + +	BARON et al. 1994; TSUKAMOTO et al. 1995; MICHIE, unpublished data
Activated α4-integrin	ND	+/–	JAKUBOWSKI et al. 1995

ND, not done; –, no effect; +/–, 1%–50% inhibition; +, 51%–60% inhibition of diabetes incidence; + +, 61%–80% inhibition; + + +, ≥80% inhibition.

demonstrated that mAbs against α4 (PS/2) and, to a lesser extent against L-selectin (MEL-14), blocked the migration of adoptively transferred splenic lymphocytes from diabetic NOD donors into islets of unmanipulated 13 week old prediabetic NOD mice. In addition, PS/2 blocked migration of splenic T cells from diabetic donors into islets of irradiated NOD mice, while MEL-14 produced only early, transient blocking.

5.2 Prevention of Insulitis and Diabetes

When α4-integrin is blocked by R1-2 or PS/2 mAb, NOD mice are significantly protected from spontaneous diabetes and adoptive transfer disease (Table 2) (YANG et al. 1993, 1994a; BARON et al. 1994; BURKLY et al. 1994; TSUKAMOTO et al. 1995; FABIEN et al. 1996). This treatment also markedly decreases the development of insulitis, suggesting that the protection from IDDM may result from direct inhibition of lymphocyte migration into the islets. Moreover, our group has shown that treatment of NOD mice with antibody against β7 integrin (mAb Fib504), α4β7 integrin (mAb DATK-32) or MAdCAM-1 (mAb MECA-367) inhibits spontaneous development of diabetes and decreases the severity of insulitis (Table 2) (YANG et al., in press; MICHIE, unpublished data).

Blockade of α4, β7 or MAdCAM-1 appears to affect neither autoimmune responses to β-cells nor immune responses to foreign antigens (YANG et al. 1993, 1994a; YANG et al., in press; BARON et al. 1994). This suggests that the IDDM-protective effect induced by blocking these AMs is not a result of immune suppression. Interestingly, we have shown that the inhibition is tissue-selective in that these treatments did not prevent salivary gland inflammation (YANG et al. 1994a; YANG et al., in press).

Two groups have examined the ability of anti-VCAM-1 antibody M/K-2.7 to inhibit the adoptive transfer of diabetes (Table 2). Baron and colleagues showed that a single dose of M/K-2.7 given at the time of cell transfer slightly decreased the degree of insulitis and marginally delayed but failed to prevent diabetes (BARON et al. 1994). In contrast, Tsukamoto et al. (1995) found that long-term treatment of host mice with M/K-2.7 reduced diabetes incidence by approximately 80%.

There are no published reports about the ability of anti-VCAM-1 treatment to prevent spontaneous diabetes. Although Tsukamoto et al. (1995) demonstrated that treatment of NOD mice with a cocktail of M/K-2.7 and anti-α4 mAb PS/2 prevented spontaneous diabetes, they did not publish data showing the effect of M/K-2.7 alone. In preliminary experiments, we have not been able to prevent the spontaneous development of diabetes by treating newborn or 8 week old NOD mice with mAbs against VCAM-1, despite achieving blocking concentrations of mAbs in the serum (Table 2) (MICHIE, unpublished data).

Jakubowski and colleagues have developed a chimeric molecule consisting of the two NH_2-terminal domains of human VCAM-1 fused to human IgG1 constant region. This chimera binds to α4 on Mn^{2+} activated murine lymphocytes but fails to bind to resting lymphocytes. In contrast, mAbs R1-2 and PS/2 bind to α4 on both resting and activated lymphocytes. Treatment of NOD mice with the chimera delays but does not prevent the adoptive transfer of diabetes, supporting a role for activated α4 in this model (Table 2) (Jakubowski et al. 1995).

6 Other Lymphocyte/Endothelial Adhesion Systems in NOD Diabetes

6.1 The Peripheral Node Adhesion System

L-selectin (CD62L) plays a major role in lymphocyte migration to PLN, and a lesser role in lymphocyte migration to PP (Gallatin et al. 1983; Hamann et al. 1994). It may also be involved in lymphocyte migration to some extranodal inflammatory sites including thymus and skin (Michie and Rouse 1991; Dawson et al. 1992; Michie et al. 1995). L-selectin's lectin domain binds to carbohydrate determinants on several endothelial glycoproteins; many of these glycoproteins react with mAb MECA-79 and are collectively known as PNAd (Streeter et al. 1988a; Baumhueter et al. 1992; Lasky et al. 1992). PNAd is highly expressed by endothelial cells in PLN but not PP.

In vitro and in vivo studies indicate the peripheral node adhesion system may play a minor role in the pathogenesis of NOD diabetes. PNAd is expressed by some peri-islet HEV in NOD pancreas (Hanninen et al. 1993; Faveeuw et al. 1994; Baumhueter et al. 1994). In contrast to MAdCAM-1, which is expressed early in inflammation, very little PNAd is seen until there is significant insulitis (approximately 8 weeks of age) (Table 1). L-selectin is expressed by some lymphocytes in inflamed islets of NOD mice (Hanninen et al. 1993, 1996; Faveeuw et al. 1994; Goldrath et al. 1995). The percentage of cells expressing L-selectin is low in the initial stages of insulitis but increases with age (Table 1).

Treatment of NOD mice with mAb Mel-14 (anti-L-selectin) leads to efficient protection against the spontaneous occurrence of IDDM if the mAb is given early during the disease process (e.g., from birth to 4 weeks of age) (Yang et al. 1994a). However, Mel-14 is unable to inhibit an ongoing diabetogenic process if administered after the onset of insulitis (Yang et al. 1994a; Lepault et al. 1995). This may simply be because most T cells in young animals have a naive phenotype (L-selectin^{+}) whereas more T cells with an activated/memory phenotype (L-selectin$^{neg/lo}$) are found in older mice. Alternatively, the antigen-specific and effector lymphocytes that are critical to development of insulitis and diabetes in older NOD mice might be found in the L-selectin^{neg} population. This alternative is supported by the finding that diabetic NOD mouse spleen cells that transfer diabetes are L-selectin^{neg} (Lepault et al. 1995).

There is conflicting data about the role of L-selectin in adoptive transfer of diabetes. Lepault and colleagues (1995) showed that pretreatment of diabetic donors with MEL-14 significantly decreased the incidence of diabetes and insulitis in the recipients. However, treatment of their host mice with MEL-14 did not block the transfer of diabetes by untreated cells. In contrast, two other groups were able to delay or prevent adoptive transfer by treating host mice with MEL-14 (YANG et al. 1993; FABIEN et al. 1996).

6.2 LFA-1 and ICAM-1

Several studies show that LFA-1 (αLβ2; CD11a/CD18) is involved in lymphocyte migration to LN, PP and numerous sites of inflammation (HAMANN et al. 1988; BARGATZE et al. 1995). Although it is unclear as to which ligands LFA-1 uses for binding to endothelium in vivo, LFA-1 ligands in vitro include Ig superfamily members ICAM-1 (CD54) and ICAM-2 (CD102) (reviewed in CARLOS and HARLAN 1994).

Several groups have shown that LFA-1 and ICAM-1 are expressed by almost all islet-infiltrating lymphocytes in NOD pancreas. ICAM-1 is also expressed by the endothelium of most islet vessels in normal and inflamed islets (HANNINEN et al. 1993; FAVEEUW et al. 1994; HASEGAWA et al. 1994; YAGI et al. 1995). In the NOD mouse, mAbs against LFA-1 (CD11a) and ICAM-1, given at a variety of ages, can prevent spontaneous diabetes. In addition, when treatment starts before 6 weeks of age, development of insulitis is also significantly inhibited (HASEGAWA et al. 1994; YAGI et al. 1995; MORIYAMA et al. 1996).

Treatment with anti-CD11a mAb combined with anti-ICAM-1 mAb is very efficient at decreasing the incidence of diabetes and insulitis in adoptive transfer models (HASEGAWA et al. 1994; YAGI et al. 1995; MORIYAMA et al. 1996). Treatment with anti-CD11a alone is less effective, while treatment with anti-ICAM-1 provides little or no protection against adoptive transfer of diabetes (BARON et al. 1994; YAGI et al. 1995; FABIEN et al. 1996).

7 The Roles of Cytokines in Pancreatic Endothelial Adhesion Molecule Expression and Inflammation

A detailed discussion of the multiple roles of cytokines in the pathogenesis of IDDM is beyond the scope of this review. However, several lines of evidence indicate cytokines may up-regulate or activate endothelial and lymphocyte AM in pancreas, facilitating the development of insulitis and subsequent destruction of beta cells. These data include: (1) inflammatory cytokines can up-regulate endothelial AM expression and lymphocyte binding in vitro; (2) cytokine activation of lymphocytes can increase the avidity of integrin AM; and (3) there are several transgenic mouse models in which expression of cytokines in islets leads to pancreatic inflammation. The inflammatory foci contain vessels that express high levels of AM including VCAM-1 (TNF-α and lymphotoxin transgenics) and MAdCAM-1 (IL-10, interferon-γ, and lymphotoxin transgenics) (Picarella et al. 1993; Wogensen et al. 1993; Lee and Sarvetnick 1994; Kratz et al. 1996). These AM are also expressed in noninflamed pancreata of lymphocyte-deficient lymphotoxin or interferon (IFN)-γ transgenics, indicating that the transgene products are responsible – either directly or indirectly – for the AM expression.

VCAM-1, MAdCAM-1 and ICAM-1 are constitutively expressed in normal pancreas and are up-regulated in NOD pancreas (Sect. 4). These molecules are known to be involved in normal migration of leukocytes from blood into tissues. Together these data suggest the following scenario of insulitis development in NOD mice: Lymphocytes use VCAM-1, MAdCAM-1 and/or ICAM-1 to migrate through pancreas during normal immunosurveillance. Autoreactive lymphocytes encounter their specific antigen in the islets, become activated, and release a variety of inflammatory mediators including cytokines and chemokines. These inflammatory mediators stimulate up-regulation of VCAM-1, MAdCAM-1 and other AM on peri-islet endothelia, thereby increasing lymphocyte recruitment. These mediators also activate leukocyte integrins, including α4β1 and α4β7, leading to strong adhesion to the endothelium and further increasing recruitment. These leukocytes, most of which have no specificity for islet antigens, enter pancreas where they amplify the inflammation and mediate β-cell destruction. Among these cytokines, TNF-α, IFN-γ, IL-10 and lymphotoxin are known to be involved in stimulation of pancreatic AM expression and to be involved in the development of insulitis (Higuchi et al. 1992; Picarella et al. 1993; Wogensen et al. 1993; Lee and Sarvetnick 1994; Yang et al. 1994b; Kratz et al. 1996). Thus, cytokines appear to play major roles in regulating the expression and function of AM that are involved in the development of IDDM.

8 α4-Integrins and Their Ligands in Human Insulin-Dependent Diabetes Mellitus

Leukocyte and endothelial AM expression has been studied in pancreata from humans with IDDM. However, such studies are hampered by the limited availability of human pancreatic tissues particularly from prediabetic or newly diagnosed diabetic individuals. In one girl with recent onset IDDM, VCAM-1 expression was found on some dendritic cells scattered throughout the pancreas but not on vascular endothelium (Hanninen et al. 1992). Another report describes VCAM-1 expression in pancreata from two people with IDDM as being identical to that seen in histologically normal pancreata (Somoza et al. 1994). There are no reports detailing the expression of α4β1, α4β7 or MAdCAM-1 in human pancreas.

9 α4-Integrins in NOD Inflamed Salivary Gland and Lacrimal Gland: Comparison to NOD Inflamed Pancreas

Besides inflammation of the pancreatic islets, NOD mice develop lymphocytic infiltration of other organs such as submandibular salivary gland and lacrimal gland (ASAMOTO et al. 1984). This makes NOD mice a valuable model to study the mechanisms involved in tissue-selective lymphocyte migration to sites of chronic inflammation. Although insulitis develops in the first few weeks of life, lymphocytic infiltration of lacrimal gland and salivary gland (sialadenitis) is not seen in our female NOD mice before 12–14 weeks of age (Michie, unpublished data). Inflamed lacrimal and salivary glands show prominent vascular expression of PNAd and VCAM-1 but no detectable expression of MAdCAM-1 (YANG et al. 1994a; FAVEEUW et al. 1994; HUNGER et al. 1996; Michie, unpublished data). The lymphocytes infiltrating the lacrimal and salivary glands are a mixed population of B and T cells that show AM expression similar to that of LN lymphocytes: LFA-1$^+$, CD44$^+$, most α4$^+$, many β7$^+$ and many L-selectin$^+$ (FAVEEUW et al. 1994; MICHIE, unpublished data). β1 expression by these cells has not been examined.

These findings suggest that α4-integrin/VCAM-1 and L-selectin/PNAd interactions may primarily mediate lymphocyte migration into lacrimal and salivary glands. In contrast, interaction of lymphocyte α4β7 with MAdCAM-1, and α4-integrins with VCAM-1, are major adhesion pathways responsible for lymphocyte migration to the islets. More importantly, tissue-selective homing to sites of chronic inflammation appears to be determined by the selective expression of vascular AM.

10 α4-Integrins in Experimental Allergic Encephalomyelitis: Comparison to NOD Diabetes

Many investigators have studied the roles of lymphocyte and endothelial AM in the pathogenesis of NOD diabetes and rodent EAE. Thus these models can be used to compare the roles of α4-integrins in two autoimmune diseases. EAE is a T cell-mediated autoimmune disease that serves as a model for human multiple sclerosis. EAE can be induced in animals from susceptible strains by immunization with central nervous system (CNS) components, by transfer of lymphocytes from an immunized animal or by transfer of specific T cell lines or clones.

ENGELHARDT provides an excellent review of α4 and VCAM-1 expression in the CNS of rodents with EAE (ENGELHARDT, this volume). Briefly, VCAM-1 is seen on a few CNS vessels in normal animals and many vessels in inflamed CNS of EAE animals (DOPP et al. 1994; STEFFEN et al. 1994; BARTEN and RUDDLE 1994; WELLER et al. 1996). One group has reported expression of MAdCAM-1 by a few vessels in spinal cords of mice with chronic relapsing EAE (O'NEILL et al. 1991).

However, vascular MAdCAM-1 expression has not been described in acute or transfer EAE (STEFFEN et al. 1994). The cells infiltrating the CNS during acute EAE are mainly $CD4^+$ T lymphocytes that are $\alpha4^+$, $\beta1^+$, $\beta7^{neg/lo}$, and L-selectin$^-$ (STEFFEN et al. 1994; WELLER et al. 1996; ENGELHARDT et al., manuscript submitted).

Functional studies in EAE indicate that α4β1/VCAM-1 interactions play a major role in migration of lymphocytes into the inflamed brain. The results are discussed in detail by ENGELHARDT (this volume) and include: (1) In vitro assays demonstrate that binding of lymphocytes to vessels in inflamed brain can be blocked by mAb against α4 or VCAM-1 but not significantly by mAb against α4β7 or β7 (YEDNOCK et al. 1992; STEFFEN et al. 1994). (2) In vivo studies show that anti-α4 or anti-VCAM-1 mAb can prevent or delay clinical disease and brain inflammation in transfer models of EAE (YEDNOCK et al. 1992; BARON et al. 1993). (3) Two studies indicate that encephalitogenic T cell clones with low levels of α4 are deficient in the ability to transfer EAE (BARON et al. 1993; KUCHROO et al. 1993).

Together, these studies indicate a major role for α4β1 and VCAM-1 in CNS inflammation in EAE. Although additional in vivo studies are needed to fully evaluate the physiologic roles of α4β7 and MAdCAM-1 in lymphocyte migration to the CNS, the current data suggest that these molecules play little role in the development of acute EAE. In contrast, all four AM are involved in the pathogenesis of NOD diabetes. A key difference between the two models is the up-regulation of endothelial MAdCAM-1 expression. MAdCAM-1 is strongly expressed in inflamed islets during all stages of insulitis in the NOD mouse. In contrast, MAdCAM-1 has only been detected in the CNS in EAE during late stages of the disease. The discordance in MAdCAM-1 expression between pancreas and CNS may be tissue-specific, with the CNS microenvironment lacking certain cells or inflammatory factors that are involved in MAdCAM-1 induction (MEBIUS et al. 1993). Alternatively, the differences in MAdCAM-1 expression may be due to strain differences in the animals used in the studies, or to the nature of the antigen and the immune response and cytokine production profile it provokes.

11 Caveats

- Studies in the NOD mouse clearly indicate that α4-integrin is involved in the pathogenesis of diabetes. However, most of these studies have failed to delineate the relative roles of α4β1 and α4β7, and their ligands VCAM-1 and MAdCAM-1, in diabetogenesis. In part, this is due to lack of specific mAbs that react with β1 or the α4β1 heterodimer. In addition, little is known about roles of different α4 activation states in the pathogenesis of IDDM. Development of mAbs, inhibitory small peptides, or AM chimeric molecules that react specifically with α4β1 vs α4β7, or with activated vs resting α4-integrins, will provide useful reagents for future studies. AM knockout mice on the NOD background might also yield useful information about the roles of α4-integrins in the development of diabetes.

- In vivo studies using mAbs against lymphocyte or endothelial AM must be carefully controlled to yield meaningful results. Sera of experimental animals must be tested to assure that adequate blocking levels of the test mAb are achieved. MAbs can cause nonspecific effects, such as destruction of lymphocytes by Fc-mediated mechanisms or by host immune response to allogeneic Ig. Thus, isotype- and species-matched mAb that binds to the same cell population as the specific mAb but does not cause changes in cell function should be used for in vivo experiments. In addition, use of Fab or F(ab)′2 fragments for in vivo studies will avoid the potential problem of Fc-mediated lymphocyte depletion. Because of the pitfalls associated with using mAbs in vivo, cells depleted of a specific population (i.e., of L-selectin$^+$ cells) can be used to examine the role of that AM in cell migration in adoptive transfer experiments.
- Although several studies have shown that mAb against α4 or its ligands can prevent insulitis and diabetes in NOD mice, the mechanisms of this inhibition are not clear. In vitro studies indicate that α4-integrins play a role in lymphocyte binding to pancreatic vessels, but only one study has directly shown that α4 is involved in lymphocyte migration from blood into islets (Faveeuw et al. 1995). Although mAbs against α4-integrins and their ligands might inhibit diabetes by directly blocking lymphocyte migration to islets, other mechanisms may also be involved. These include interruption of priming or activation of autoreactive T cells, mAb-mediated deletion or activation of lymphocyte subsets, deviation of the predominant immune response to islet antigens from Th1 to Th2, and alteration of cytolytic interactions between effector leukocytes and islet β-cells. These possible mechanisms need to be investigated using in vitro and in vivo assays.
- Migration of lymphocytes from blood to pancreatic islets is almost certainly a multistep process involving sequential adhesion and activation events. Thus, interruption of any single step in the cascade would be expected to inhibit the lymphocyte migration. This has positive implications for the development of anti-AM based therapies in humans. However, it makes it difficult to assess the roles of various AM in lymphocyte migration to NOD islets. In vivo microscopy, in which islet vessels are directly observed for ability of anti-AM mAbs to block various steps of the adhesion cascade, might be useful in determining the dominant adhesion mechanisms involved in lymphocyte migration to islets (Bargatze et al. 1995).
- There may be significant differences in immunologic mechanisms, including utilization of lymphocyte/endothelial adhesion pathways, in the pathogenesis of spontaneous and adoptive transfer diabetes in NOD mice (Lo et al. 1993). In addition, adoptive transfer models may vary significantly from one investigator to the next, especially in the age and strain (NOD vs NOD-SCID) of the host.
- Although the immunologic mechanisms that cause diabetes in NOD mice are thought to be similar to those of human IDDM, no inbred animal is a perfect

model for human disease. Experimental results indicate α4-integrins are involved in diabetogenesis in NOD mice. However, these results must be followed up by in vitro studies using human pancreas and lymphocytes before anti-AM based therapies can be tried in humans.

12 Summary

Lymphocyte/endothelial adhesion followed by transendothelial migration is a key event in the development of organ-specific autoimmunity. Selective interactions of cell surface AM regulate lymphocyte migration under normal as well as pathologic inflammatory conditions. NOD mice are an ideal model for investigating the roles of AM in regulation of lymphocyte migration to target organs in autoimmune diseases such as IDDM. Both in vitro and in vivo studies in NOD mice strongly suggest that the mucosal (α4β7/MAdCAM-1) adhesion system and α4-integrin/VCAM-1 appear to be prominent pathways for insulitis development. In contrast, α4-mediated interactions in NOD inflamed salivary and lacrimal gland and in the inflamed CNS of rodents with EAE seem to be dominated by α4-integrins and VCAM-1. The fact that blocking α4-integrin pathways in NOD mice leads to successful interruption of the diabetogenic process suggests that AM provide a potential therapeutic target for human IDDM. Further studies on IDDM patients will prove helpful for understanding IDDM pathogenesis and in providing a basis for designing AM-based therapeutic approaches.

Acknowledgements. This work was supported by grants from the Department of Veterans Affairs (VA) Office of Research and Development, the American Diabetes Association and the National Institutes of Diabetes and Digestive and Kidney Diseases. Huey-Kang Sytwu was supported by a fellowship from the National Defense Medical Center, Taipei, Taiwan, Republic of China and in part by a grant from the National Institutes of Health (CA 49734). We thank Drs. Britta Engelhardt, Gregory Johnson and Leslie McEvoy for thoughtful review of this manuscript.

References

Asamoto H, Akazawa Y, Tashiro S, Oishi M, Azuma T, Koide S, Sudo K, Yokota H, Tochino Y (1984) Infiltration of lymphocytes in various organs of the NOD (non-obese diabetic) mouse. J Jpn Diabetic Soc 27:775–781

Bach J-F (1994) Insulin-dependent diabetes mellitus as an autoimmune disease. Endocr Rev 15:516–542

Bargatze RF, Jutila MA, Butcher EC (1995) Distinct roles of L-selectin and α4β7 and LFA-1 in lymphocyte homing to Peyer's patch HEV in situ: the multistep model confirmed and refined. Immunity 3:99–108

Baron JL, Madri JA, Ruddle NH, Hashim G, Janeway CA Jr (1993) Surface expression of α4 integrin by CD4+ T cells is required for their entry into brain parenchyma. J Exp Med 177:57–68

Baron JL, Reich EP, Visintin I, Janeway CA Jr (1994) The pathogenesis of adoptive murine autoimmune diabetes requires an interaction between alpha 4-integrins and vascular cell adhesion molecule-1. J Clin Invest 93:1700–1708

Barten DM, Ruddle NH (1994) Vascular cell adhesion molecule-1 modulation by tumor necrosis factor in experimental allergic encephalomyelitis. J Neuroimmunol 51:123–133

Baumhueter S, Singer MS, Henzel W, Hemmerich S, Renz M, Rosen SD, Lasky LA (1992) Binding of L-selectin to the vascular sialomucin CD34. Science 262:436–438

Baumhueter S, Dybdal N, Kyle C, Lasky LA (1994) Global vascular expression of murine CD34, a sialomucin-like endothelial ligand for L-selectin. Blood 84(8):2554–2565

Berlin C, Berg EL, Briskin MJ, Andrew DP, Kilshaw PJ, Holzmann B, Weissman IL, Hamann A, Butcher EC (1993) $\alpha 4\beta 7$ integrin mediates lymphocyte binding to the mucosal vascular addressin MAdCAM-1. Cell 74:185–195

Briskin MJ, McEvoy L, Butcher EC (1993) The mucosal vascular addressin, MAdCAM-1, displays homology to immunoglobulin and mucin-like adhesion receptors and to IgA. Nature 363:461–464

Briskin MJ, Winsor-Hines D, Shyjan A, Bloom S, Wilson J, McEvoy LM, Butcher EC, Kassam N, MacKay CR, Newman W, Ringler DJ. Expression of human mucosal addressin molecule is restricted to intestinal tract and associated lymphoid tissue (submitted)

Burkly LC, Jakubowski A, Hattori M (1994) Protection against adoptive transfer of autoimmune diabetes mediated through very late antigen-4 integrin. Diabetes 43:529–534

Butcher EC (1991) Leukocyte-endothelial cell recognition: three (or more) steps to specificity and diversity. Cell 67:1033–1036

Butcher EC, Picker LJ (1996) Lymphocyte homing and homeostasis. Science 272:60–66

Carlos TM, Harlan JM (1994) Leukocyte-endothelial adhesion molecules. Blood 84:2068–2101

Castano L, Eisenbarth GS (1990) Type-I diabetes: a chronic autoimmune disease of human, mouse, and rat. Annu Rev Immunol 8:647–679

Chan BCM, Elices MJ, Murphy E, Hemler ME (1992) Adhesion to vascular cell adhesion molecule 1 and fibronectin. Comparison of $\alpha 4\beta 1$ (VLA-4) and $\alpha 4\beta 7$ on the human B cell line JY. J Biol Chem 267:8366–8370

Dawson J, Sedgwick AD, Edwards JCW, Lees P (1992) The monoclonal antibody MEL-14 can block lymphocyte migration into a site of chronic inflammation. Eur J Immunol 22:1647–1650

Dopp JM, Breneman SM, Olschowka JA (1994) Expression of ICAM-1, VCAM-1, L-selectin, and leukosialin in the mouse central nervous system during the induction and remission stages of experimental allergic encephalomyelitis. J Neuroimmunol 54:129–144

Elices MJ, Osborn L, Takada Y, Crouse C, Luhowskyj S, Hemler ME, Lobb RR (1990) VCAM-1 on activated endothelium interacts with the leukocyte integrin VLA-4 at a site distinct from the VLA-4/fibronectin binding site. Cell 60:577–584

Engelhardt B, Martin-Simonet MTG, Rott L, Butcher EC, Michie SA. T lymphocytes present in sites of chronic inflammation can be distinguished by their expression of tissue-selective cell adhesion molecules (submitted)

Fabien N, Bergerot I, Orgiazzi J, Thivolet C (1996) Lymphocyte function associated antigen-1, integrin alpha 4, and L-selectin mediate T-cell homing to the pancreas in the model of adoptive transfer of diabetes in NOD mice. Diabetes 45:1181–1186

Faveeuw C, Gagnerault MC, Lepault F (1994) Expression of homing and adhesion molecules in infiltrated islets of Langerhans and salivary glands of nonobese diabetic mice. J Immunol 152:5969–5978

Faveeuw C, Gagnerault MC, Kraal G, Lepault F (1995) Homing of lymphocytes into islets of Langerhans in prediabetic non-obese diabetic mice is not restricted to autoreactive T cells. Int Immunol 7:1905–1913

Fries JWU, Williams AJ, Atkins RC, Newman W, Lipscomb MF, Collins T (1993) Expression of VCAM-1 and E-selectin in an in vivo model of endothelial activation. Am J Pathol 143:725–737

Gallatin WM, Weissman IL, Butcher EC (1983) A cell-surface molecule involved in organ-specific homing of lymphocytes. Nature 304:30–34

Goldrath AW, Barber L, Chen KE, Alters SE (1995) Differences in adhesion markers, activation markers, and TCR in islet infiltrating vs peripheral lymphocytes in the NOD mouse. J Autoimmunity 8: 209–220

Gowans JL, Knight EJ (1964) The route of recirculation of lymphocytes in the rat. Proc R Soc Lond Biol Sci 159:257–282

Hamann A, Jablonski-Westrich D, Duijvestijn A, Butcher EC, Baisch H, Harder R, Thiele H-G (1988) Evidence for an accessory role of LFA-1 in lymphocyte-high endothelium interaction during homing. J Immunol 140:693–699

Hamann A, Andrew DP, Jablonski-Westrich D, Holzmann B, Butcher EC (1994) Role of $\alpha 4$-integrins in lymphocyte homing to mucosal tissues in vivo. J Immunol 152:3282–3293

Hanninen A, Jalkanen S, Salmi M, Toikkanen S, Nikolakaros G, Simell O (1992) Macrophages, T cell receptor usage, and endothelial cell activation in the pancreas at the onset of insulin-dependent diabetes mellitus. J Clin Invest 90:1901–1910

Hanninen A, Taylor C, Streeter PR, Stark LS, Sarte JM, Shizuru JA, Simell O, Michie SA (1993) Vascular addressins are induced on islet vessels during insulitis in nonobese diabetic mice and are involved in lymphoid cell binding to islet endothelium. J Clin Invest 92:2509–2515

Hanninen A, Salmi M, Simell O, Jalkanen S (1996) Mucosa-associated (beta 7-integrin high) lymphocytes accumulate early in the pancreas of NOD mice and show aberrant recirculation behavior. Diabetes 45:1173–1180

Hasegawa Y, Yokono K, Taki T, Amano K, Tominaga Y, Yoneda R, Yagi N, Maeda S, Yagita H, Okumua K, Kasuga M (1994) Prevention of autoimmune insulin-dependent diabetes in non-obese diabetic mice by anti-LFA-1 and anti-ICAM-1 mAb. Int Immunol 6:831–838

Hesterberg PE, Winsor-Hines D, Briskin MJ, Soler-Ferran D, Merrill C, Mackay CR, Newman W, Ringler DJ (1996) Rapid resolution of chronic colitis in the cotton-top tamarin with an antibody to a gut-homing integrin alpha4 beta7. Gastroenterology 111:1373–1380

Higuchi Y, Herrera P, Muniesa PJH, Belin D, Ohashi P, Aichele P, Orci L, Vassalli J-D, Vassalli P (1992) Expression of a tumor necrosis factor α transgene in murine pancreatic β cells results in severe and permanent insulitis without evolution towards diabetes. J Exp Med 176:1719–1731

Holzmann B, McIntyre BW, Weissman IL: (1989) Identification of a murine Peyer's patch specific lymphocyte homing receptor as an integrin molecule with an α chain homologous to human VLA-4α. Cell 56:37–46

Hu MCT, Crowe DT, Weissman IL, Holzmann B (1992) Cloning and expression of mouse integrin βp(β7): a functional role in Peyer's patch-specific lymphocyte homing. Proc Natl Acad Sci USA 89:8254–8258

Hunger RE, Muller S, Laissue JA, Hess MW, Carnaud C, Garcia I, Mueller C (1996) Inhibition of submandibular and lacrimal gland infiltration in nonobese diabetic mice by transgenic expression of soluble TNF-receptor p55. J Clin Invest 98:954–961

Hynes RO (1992) Integrins: versatility, modulation, and signaling in cell adhesion. Cell 69:11–25

Imhof BA, Dunon D (1995) Leukocyte migration and adhesion. Adv Immunol 58:345–416

Jacobsen K, Kravitz J, Kincade PW et al (1996) Adhesion receptors on bone marrow stromal cells: in vivo expression of vascular cell adhesion molecule-1 by reticular cells and sinusoidal endothelium in normal and gamma-irradiated mice. Blood 87:73–82

Jakubowski A, Ehrenfels BN, Pepinsky RB, Burkly LC (1995) Vascular cell adhesion molecule-Ig fusion protein selectively targets activated alpha 4-integrin receptors in vivo. Inhibition of autoimmune diabetes in an adoptive transfer model in nonobese diabetic mice. J Immunol 155:938–946

Kilshaw PJ, Murant SJ (1991) Expression and regulation of β7 (βp) integrins on mouse lymphocytes: relevance to the mucosal immune system. Eur J Immunol 21:2591–2597

Koch AE, Burrows JC, Haines GK, Carlos TM, Harlan JM, Leibovich SJ (1991) Immunolocalization of endothelial and leukocyte adhesion molecules in human rheumatoid and osteoarthritic synovial tissues. Lab Invest 64:313–320

Kraal G, Schornagel K, Streeter PR et al (1995) Expression of the mucosal vascular addressin, MAdCAM-1, on sinus-lining cells in the spleen. Am J Pathol 147:763–771

Kratz A, Campos-Neto A, Hanson MS, Ruddle NH (1996) Chronic inflammation caused by lymphotoxin is lymphoid neogenesis. J Exp Med 183:1461–1472

Kuchroo VK, Martin CA, Greer JM, Ju ST, Sobel RA, Dorf ME (1993) Cytokines and adhesion molecules contribute to the ability of myelin proteolipid protein-specific T cell clones to mediate experimental allergic encephalomyelitis. J Immunol 151:4371–4382

Lasky LA, Singer MS, Dowbenko D, Imai Y, Henzel WJ, Grimley C, Fennie C, Gillett N, Watson SR, Rosen SD (1992) An endothelial ligand for L-selectin is a novel mucin-like molecule. Cell 69:927–938

Lee MS, Sarvetnick N (1994) Induction of vascular addressins and adhesion molecules in the pancreas of IFN-gamma transgenic mice. J Immunol 152:4597–4603

Lepault F, Gagnerault MC, Faveeuw C, Bazin H, Boitard C (1995) Lack of L-selectin expression by cells transferring diabetes in NOD mice: insights into the mechanisms involved in diabetes prevention by MEL-14 antibody treatment. Eur J Immunol 25:1502–1507

Lobb RR, Hemler ME (1994) The pathophysiologic role of a4-Integrins in vivo. J Clin Invest 94: 1722–1728

Lo D, Reilly CR, Scott B, Liblau R, McDevitt HO, Burkly LC (1993) Antigen-presenting cells in adoptively transferred and spontaneous autoimmune diabetes. Eur J Immunol 23:1693–1698

Mackay CR, Marston WL, Dudler L (1990) Naive and memory T cells show distinct pathways of lymphocyte recirculation. J Exp Med 171:801–817

Mebius RE, Breve J, Kraal G, Streeter PR (1993) Developmental regulation of vascular addressin expression: a possible role for site-associated environments. Int Immunol 5:443–449

Michie SA, Rouse RV (1991) Traffic of peripheral B and T lymphocytes to hyperplastic, preneoplastic thymuses of AKR mice. Am J Pathol 138:1015–1025

Michie SA, Streeter PR, Butcher EC, Rouse RV (1995) L-selectin and α4β7 integrin homing receptor pathways mediate peripheral lymphocyte traffic to AKR mouse hyperplastic thymus. Am J Pathol 147:412–421

Miyake K, Medina K, Ishihara K, Kimoto M, Auerbach R, Kincade PW (1991) A VCAM-like adhesion molecule on murine bone marrow stromal cells mediates binding of lymphocyte precursors in culture. J Cell Biol 114:557–565

Moriyama H, Yokono K, Amano K, Nagata M, Hasegawa Y, Okamoto N, Tsukamoto K, Miki M, Yoneda R, Yagi N et al (1996) Induction of tolerance in murine autoimmune diabetes by transient blockade of leukocyte function-associated antigen-1/intercellular adhesion molecule-1 pathway. J Immunol 157:3737–3743

O'Neill JK, Butter C, Baker D, Gschmeissner SE, Kraal G, Butcher EC, Turk JL (1991) Expression of vascular addressins and ICAM-1 by endothelial cells in the spinal cord during chronic relapsing experimental allergic encephalomyelitis in the Biozzi AB/H mouse. Immunology 72:520–525

Picarella DE, Kratz A, Li CB, Ruddle NH, Flavell RA (1993) Transgenic tumor necrosis factor (TNF)-alpha production in pancreatic islets leads to insulitis, not diabetes. J Immunol 150:4136–4150

Picarella D, Hurlbut P, Rottman J, Shi X, Butcher E, Ringler DJ (1997) Monoclonal antibodies specific for β7 integrin and MAdCAM-1 reduce inflammation in the colon of scid mice reconstituted with CD45RBhi CD4+ T cells. J Immunol 158:2099–2106

Postigo AA, Teixido J, Sanchez-Madrid F (1993) The alpha 4 beta 1/VCAM-1 adhesion pathway in physiology and disease. Res Immunol 144:723–735

Rice GE, Munro JM, Corless C, Bevilacqua MP (1991) Vascular and nonvascular expression of INCAM-110. A target for mononuclear leukocyte adhesion in normal and inflammed human tissues. Am J Pathol 138:385–393

Rossini AA, Greiner DL, Friedmen HP, Mordes JP (1993) Immunopathogenesis of diabetes mellitus. Diabetes Rev 1:43–75

Ruegg C, Postigo AA, Sikorski EE, Butcher EC, Pytela R, Erle DJ (1992) Role of integrin alpha 4/beta P in lymphocyte adherence to fibronectin and VCAM-1 and in homotypic cell clustering. J Cell Biol 117:179–189

Shimizu Y, Newman W, Tanaka Y, Shaw S (1992) Lymphocyte interactions with endothelial cells. Immunol Today 13:106–112

Sikorski EE, Hallmann R, Berg EL, Butcher EC (1993) The Peyer's patch high endothelial receptor for lymphocytes, the mucosal vascular addressin, is induced on a murine endothelial cell line by TNF-α and IL-1. J Immunol 151:5239–5250

Somoza N, Vargas F, Roura-Mir C, Vives-Pi M, Fernandez-Figueras MT, Ariza A, Gomis R, Bragado R, Marti M, Jaraquemada D, Pujol-Borrell R (1994) Pancreas in recent onset insulin-dependent diabetes mellitus. Changes in HLA, adhesion molecules and autoantigens, restricted T cell receptor Vβ usage, and cytokine profile. J Immunol 153:1360–1377

Springer TA (1994) Traffic signals for lymphocyte recirculation and leukocyte emigration: the multistep paradigm. Cell 76:301–314

Stamper HB Jr, Woodruff JJ (1976) Lymphocyte homing into lymph nodes. In vitro demonstration of the selective affinity of recirculating lymphocytes for high-endothelial venules. J Exp Med 144:828–833

Steffen BJ, Butcher EC, Engelhardt B (1994) Evidence for involvement of ICAM-1 and VCAM-1 in lymphocyte interaction with endothelium in experimental autoimmune encephalomyelitis in the central nervous system in the SJL/J mouse. Am J Pathol 145:189–201

Streeter PR, Rouse BT, Butcher EC (1988a) Immunohistologic and functional characterization of a vascular addressin involved in lymphocyte homing into peripheral lymph nodes. J Cell Biol 107:1853–1862

Streeter PR, Berg EL, Rouse BTW, Bargatze RK, Butcher EC (1988b) A tissue specific endothelial cell molecule involved in lymphocyte homing. Nature 331:41–46

Szabo MC, Butcher EC, McEvoy LM (1997) Specialization of mucosal follicular dendritic cells revealed by mucosal addressin-cell adhesion molecule-1 display. J Immunol 158:5584–5588

Tsukamoto K, Yokono K, Amano K, Nagata M, Yagi N, Tominaga Y, Moriyama H, Miki M, Okamoto N, Yoneda R et al (1995) Administration of monoclonal antibodies against vascular cell adhesion molecule-1/very late antigen-4 abrogates predisposing autoimmune diabetes in NOD mice. Cell Immunol 165:193–201

Wayner EA, Garcia-Pardo A, Humphries MJ, McDonald JA, Carter WG (1989) Identification and characterization of the T lymphocyte adhesion receptor for an alternative cell attachment domain (CS-1) in plasma fibronectin. J Cell Biol 109:1321–1330

Weller RO, Engelhardt B, Phillips MJ (1996) Lymphocyte targeting of the central nervous system: a review of afferent and efferent CNS-immune pathways. Brain Pathol 6:275–288

Wogensen L, Huang X, Sarvetnick N (1993) Leukocyte extravasation into the pancreatic tissue in transgenic mice expressing interleukin 10 in the islets of Langerhans. J Exp Med 178:175–185

Yagi N, Yokono K, Amano K, Nagata M, Tsukamoto K, Hasegawa Y, Yoneda R, Okamoto N, Moriyama H, Miki M et al (1995) Expression of intercellular adhesion molecule 1 on pancreatic beta-cells accelerates beta-cell destruction by cytotoxic T-cells in murine autoimmune diabetes. Diabetes 44:744–752

Yang X-D, Karin N, Tisch R, Steinman L, McDevitt HO (1993) Inhibition of insulitis and prevention of diabetes in nonobese diabetic mice by blocking L-selectin and very late antigen-4 adhesion receptors. Proc Natl Acad Sci USA 90:10494–10498

Yang X-D, Michie SA, Karin N, Tisch R, Steinman L, McDevitt HO (1994a) A predominant role of integrin α4 in the spontaneous development of autoimmune diabetes in nonobese diabetic mice. Proc Natl Acad Sci USA 91:12604–12608

Yang X-D, Tisch R, Singer SM, Cao ZA, Liblau R, Schreiber RD, McDevitt HO (1994b) Effect of tumor necrosis factor α on insulin-dependent diabetes mellitus in NOD mice. I. The early development of autoimmunity and diabetogenic process. J Exp Med 180:995–1004

Yang X-D, Sytwu H-K, McDevitt HO, Michie SA. Involvement of beta7 integrin and mucosal addressin cell adhesion molecule-1 (MAdCAM-1) in the development of diabetes in nonobese diabetic mice. Diabetes (in press)

Yednock TA, Cannon C, Fritz LC, Sanchez-Madrid F, Steinman L, Karin N (1992) Prevention of experimental autoimmune encephalomyelitis by antibodies against α4β1 integrin. Nature 356:63–66

Immunosuppression by Blocking α4-Integrins/VCAM-1 Adhesion

M. Isobe[1], J. Hori[2], and J. Suzuki[1]

1 Introduction . . . 85
2 Induction of VCAM-1 on Rejecting Cardiac Myocytes . . . 86
3 Effects of Antibodies to VLA-4 and VCAM-1 to Cardiac Allografts . . . 87
4 Cardiac Allograft Survival and Skin Grafts . . . 88
5 Characterization of Immune Suppression . . . 89
6 Effects of Anti-VLA-4 Monoclonal Antibodies on Cornea and Skin Allografts . . . 90
7 Unresponsiveness to Soluble Antigens . . . 92
8 Regulation of Immunosuppression by Cytokines . . . 93
9 Effects on Chronic Rejection of Cardiac Allografts . . . 93
10 Conclusion . . . 95
References . . . 95

1 Introduction

Allograft rejection is the expression of an immune reaction of the recipient against foreign antigens of the transplanted tissue and is mostly mediated by T lymphocytes. The initial step consists of T lymphocyte adhesion to graft endothelium, a process that is mediated by cell adhesion molecules. It has been shown that there are several pathways for recognition of alloantigens on the transplanted vascularized organs, and cell adhesion molecules play important roles in these processes. Adhesion is necessary for tightening of contact between T cells and antigen-presenting cells. Two signals are required for optimal clonal expansion of T cells: one from the T cell receptor upon binding to the major histocompatibility complex (MHC) antigen and the other signal from receptors that are distinctive from T cell receptors (Mueller et al. 1989; Schwartz et al. 1989; Schwartz 1990). Shaw et al. found that several adhesion molecules produce such costimulatory signals

[1]The First Department of Internal Medicine, Shinshu University School of Medicine, 3-1-1 Asahi, Matsumoto 390, Japan
[2]Department of Ophthalmology, University of Tokyo, School of Medicine, 7-3-1 Hongo, Bunkyo-ku, Tokyo 113, Japan

(Clevers et al. 1988; Shimizu et al. 1990a; Weaver and Unaue 1990). Schwartz et al. showed that, in the absence of the costimulatory signal, T cells become inactivated against the particular antigen, which leads to clonal anergy of the T cells (Mueller et al. 1989).

LFA-1, CD28 and VLA-4 are molecules which generate costimulatory signals (Springer et al. 1987; Freeman et al. 1989; Simmons et al. 1989; Osborn et al. 1989; van Seventer et al. 1990, 1991; Shimizu et al. 1990b; Springer 1990; Damle and Aruffo 1991; Jenkins et al. 1991; Koulova et al. 1991; Linsley et al. 1991; Moingeon et al. 1991; Kato et al. 1992; Damle et al. 1992; Harding et al. 1992; Azuma et al. 1993). Numerous experimental and clinical studies have confirmed the induction of adhesion molecules on endothelial cells in rejecting organs (Allen et al. 1992; Steinhoff et al. 1993) and have demonstrated reduced severity of allograft rejection when the expression of adhesion molecules is effectively blocked (Cobbold et al. 1989; Qin et al. 1989; Isobe et al. 1992b; Lenschow et al. 1992; Chavin et al. 1993; Nakakura et al. 1993; Nicolls et al. 1993; Wood et al. 1993; Isobe et al. 1994; Larsen et al. 1996). We reported that simultaneous administration of monoclonal antibodies (mAbs) to ICAM-1 and LFA-1 results in an indefinite acceptance of cardiac (Isobe and Ihara 1993; Isobe et al. 1992b) and primary skin allografts (Isobe et al. 1996). Cardiac allografts acceptance is always accompanied by specific acceptance of secondary skin allografts, indicating that tolerance is induced.

The roles of VCAM-1/VLA-4 adhesion in the immune reaction have been well characterized. However, how adhesion between these molecules contributes to allograft rejection remains poorly understood. Therefore, we investigated the roles of these molecules in allograft rejection of heart (Isobe et al. 1994), and cornea (Hori et al. 1996) and the immune response to soluble antigens (Isobe et al. 1994).

2 Induction of VCAM-1 on Rejecting Cardiac Myocytes

Expression of VCAM-1 and ICAM-1 in normal cardiac tissue is very limited in vascular endothelium. However, immunohistochemical studies on transplanted allografts revealed that these molecules are greatly induced on the vascular endothelium and on cardiac myocytes from the early stage of graft rejection in human transplants or animal models of transplantation (Cosimi et al. 1990; Taylor et al. 1992; Orosz et al. 1993; Pelletier et al. 1993; Tanio et al. 1994). Therefore, it appears that VCAM-1/VLA-4 or ICAM-1/LFA-1 interaction participates in the pathophysiology of allograft rejection.

In our experiment we transplanted BALB/c mouse heart into C3H/He recipients. As shown in Fig. 1, expression of VCAM-1 was greatly enhanced not only on the endothelial cells but also on the cardiac myocytes of untreated rejecting allografts. This enhancement is followed by development of myocyte necrosis.

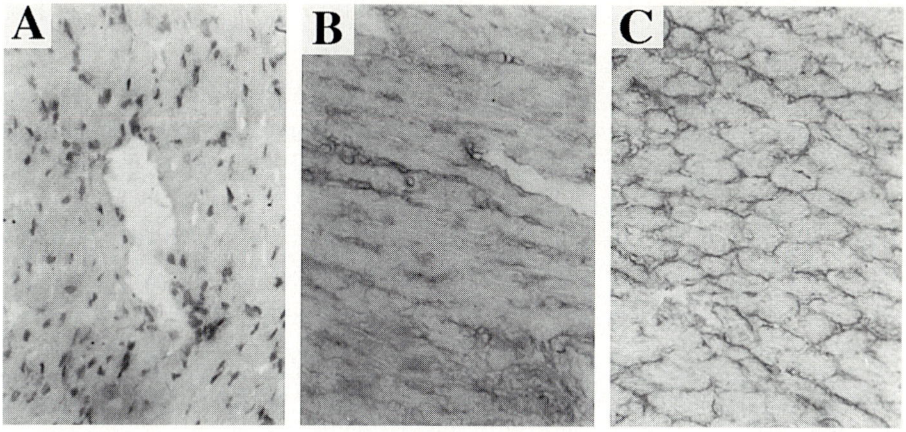

Fig. 1A–C. Immunohistopathology of cardiac allografts stained with anti-VCAM-1 mAb. **A** Normal heart; vascular endothelial cells are faintly stained. **B** An allograft without treatment sacrificed at 5 days and **C** 8 days after transplantation. Marked induction of VCAM-1 on cardiac myocytes as well as the vascular wall is noted. ×200

VCAM-1 expression on certain antigen-presenting cells (APCs) may play an important role in activation of $CD4^+$ T cells by mediating the binding to, and costimulating proliferation of, antigen-specific $VLA\text{-}4^+$ T cells (Damle and Aruffo 1991; van Seventer et al. 1991). It is likely that the up-regulation of VCAM-1 is crucial in eliciting rejection by recruiting T cells and macrophages into the inflamed tissues. These observation regarding the role of VLA-4 and VCAM-1 interaction prompted us to test whether blockade of VLA-4/VCAM-1 adhesion may allow induction of specific immunosuppression similar to that observed with mAbs to ICAM-1/LFA-1.

3 Effects of Antibodies to VLA-4 and VCAM-1 to Cardiac Allografts

An outline of the experiment is shown in Fig. 2. Donor hearts were heterotopically transplanted into recipients using a microsurgical technique (Ono and Lindsey 1969; Isobe et al. 1991, 1992a, b). C3H/He hearts were transplanted into same strain recipients as isograft controls. BALB/c hearts were transplanted into C3H/He recipients as allografts. Recipients received 100 μg daily of anti-VCAM-1 or anti-VLA-1 mAbs, or both for 5 consecutive days starting immediately after transplantation. The graft beat was checked daily by palpation. The complete cessation of graft beat was interpreted as rejection.

Approximately 1 cm^2 of mouse back skin was transplanted onto the recipient's back as shown previously (Isobe et al. 1994). Observation was made by two in-

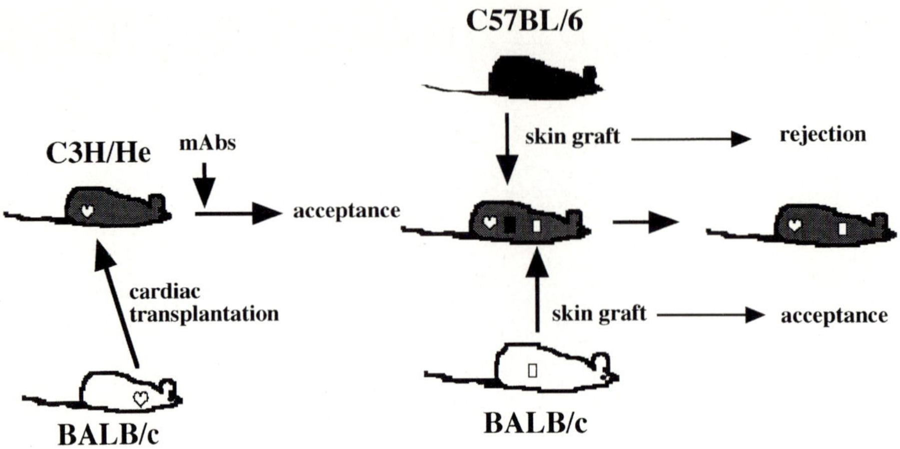

Fig. 2. Outline of the animal experiment

dependent examiners, one of them unaware of the treatment. Complete loss of graft tissue was interpreted as rejection.

Hybridomas producing M/K-2 (rat IgG1) (Miyake et al. 1991a; Hession et al. 1992) and PS/2 (rat IgG2b) (Miyake et al. 1991b) mAbs that react with murine VCAM-1 and VLA-4, respectively, were gift from Professor K. Okumura of Juntendo University. M18/2 (anti-CD18, rat IgG2a), also obtained from Prof. Okumura, was used as a control mAb because M18/2 does not block cell-mediated target cell lysis in vitro (Sanchez et al. 1983).

4 Cardiac Allograft Survival and Skin Grafts

As shown in Fig. 3, graft survival in mice treated with M/K-2 (anti-VCAM-1) (median survival 20 days) and those treated with PS/2 (anti-VLA-4) (30 days) was greater than in control mice (8 days), in which cardiac allografts were rejected within 10 days. Cardiac allografts treated with 100 μg each of anti-VCAM-1 and anti-VLA-1 mAbs kept beating longer than allografts in mice without treatment. Eight of 18 mice treated with both M/K-2 and PS/2 accepted the grafts over 65 days and five of them accepted the grafts over 100 days.

Mice with long-surviving cardiac grafts were challenged with skin grafts from donor (BALB/c) and third-party (C57BL/6) strains. Survival of the donor type skin was significantly greater than that of third-party skin (Fig. 4). One mouse accepted a skin allograft indefinitely (Fig. 5). These results indicate that in vivo administration of anti-VCAM-1 and anti-VLA-4 mAbs induces specific immunological unresponsiveness to cardiac allografts.

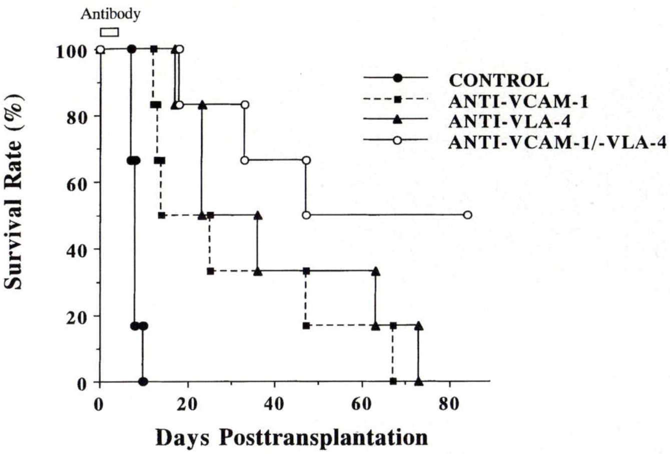

Fig. 3. Survival rate of cardiac allografts after treatment with monoclonal antibodies. Recipient mice were injected with either 100 μg of control antibody, anti-VCAM-1, anti-VLA-4, or 50 μg each of anti-VCAM-1 and anti-VLA-4 starting just after the operation and lasting until day 5 of transplantation

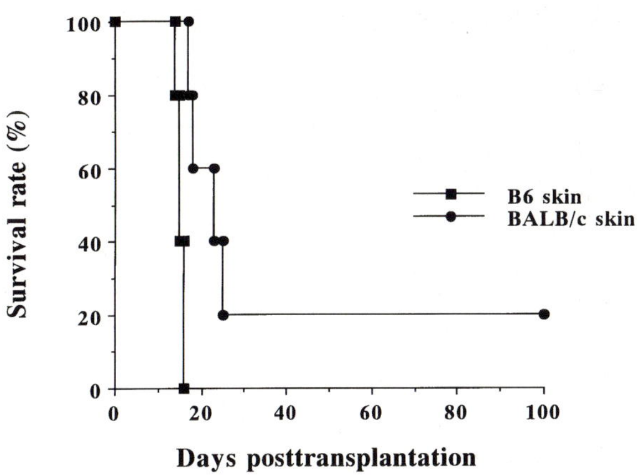

Fig. 4. Survival rate of second skin grafts. Survival of second skin grafts from the donor strain was significantly longer than that from the third-party strain

5 Characterization of Immune Suppression

Indirect immunofluorescent staining was performed using PS/2 as a primary antibody. Direct staining was also performed using FITC-conjugated PS/2 mAb. The cells were then examined by FACS analysis. Splenocytes from PS/2 treated mice showed almost complete blockade of VLA-4 molecules 7 days after mAb treatment. Transplantation of cardiac allograft did not alter the expression of VLA-4

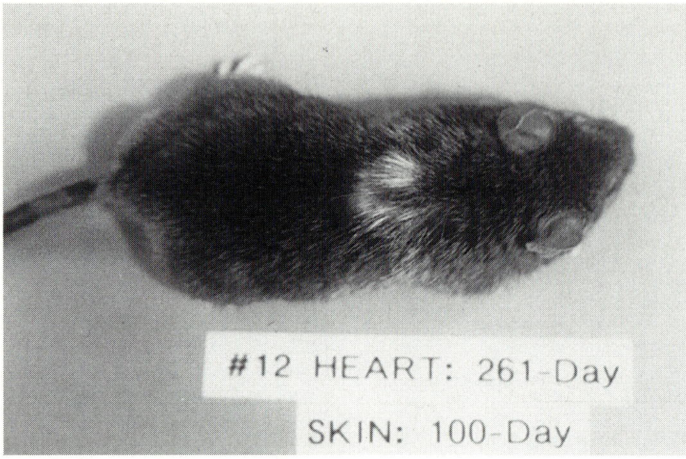

Fig. 5. A mouse with a cardiac allograft and 5 day course of monoclonal antibodies to VCAM-1 and VLA-4, showing acceptance of the second skin graft from the donor strain

molecule on the splenocytes. VLA-4 molecules on these splenocytes were saturated with PS/2 mAb administered in vivo. VLA-4 expression on splenocytes of allografted mice indicated that the mAb treatment led to significant reduction of VLA-4-positive cells 7 days after transplantation. VLA-4 expression recovered to the normal level at 50 days after mAb treatment.

It was of interest to know whether immunosuppression by mAbs to T cell surface antigen accompanied T cell depletion and changes in T cell subset. A 6 day course of anti-VCAM-1/anti-VLA-4 treatment of ungrafted C3H/He mice did not reduce the circulating leukocyte count or yield of leukocytes per spleen. The slight increase in white blood cell count after mAb treatment, probably due to major alterations in lymphocyte traffic and/or distribution, was observed (Issekutz 1991). At day 7, $CD4^+$ and $CD8^+$ subpopulations of splenocytes were not reduced by the treatment (Fig. 6).

6 Effects of Anti-VLA-4 Monoclonal Antibodies on Cornea and Skin Allograft

Effects of mAbs on VLA-4 and LFA-1 were evaluated in our experimental model of corneal (Hori et al. 1996) and skin (unpublished data) transplantation. C3H/He donor corneas were transplanted into BALB/c corneal beds. Fourteen of 16 allografts in nontreated mice and control mAb-treated mice became opaque by 2 weeks after transplantation. The allografts treated with anti-VLA-4 or anti-LFA-1 mAb alone, or with both mAbs, remained transparent for more than 2 weeks. Although

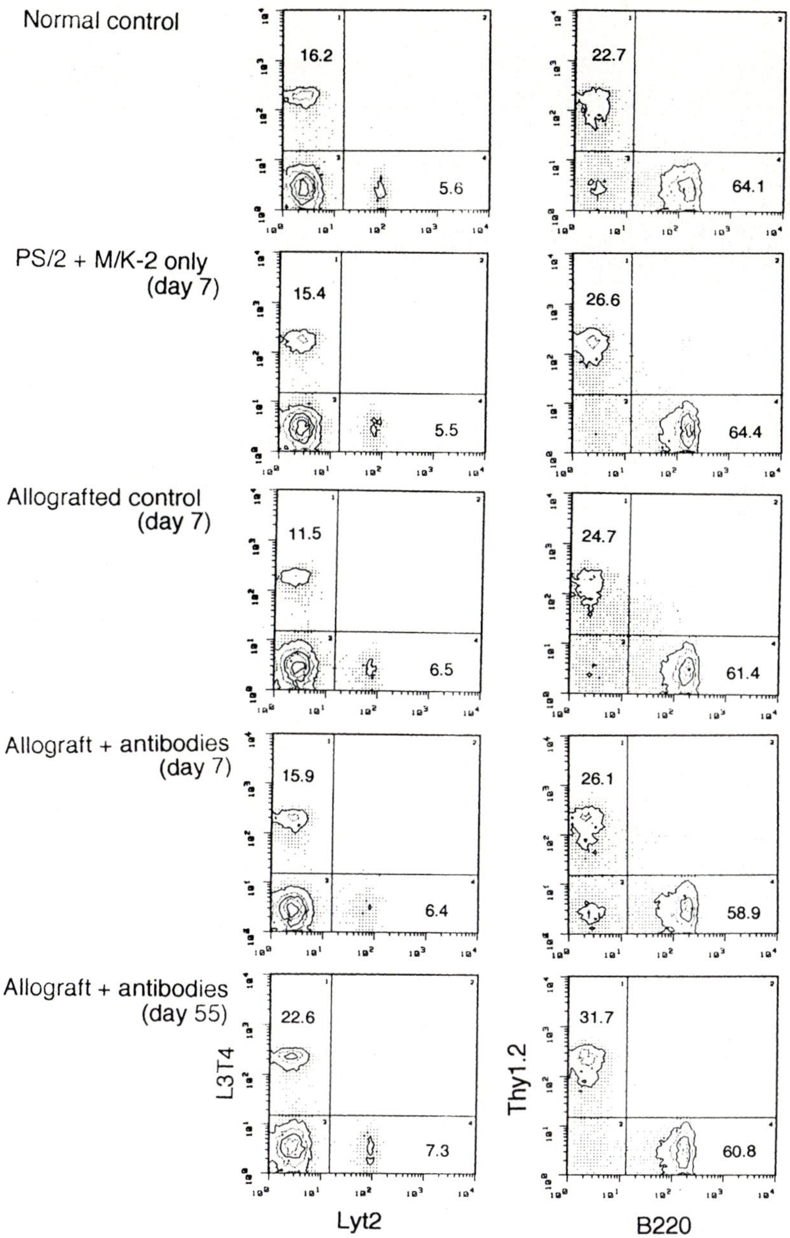

Fig. 6. CD4 and CD8 expression on splenocytes from recipient mice treated with anti-VLA-4 and anti-VCAM-1 monoclonal antibodies 8 days after transplantation were analyzed by flow cytometry. No significant changes in CD4/CD8 subset were observed

Table 1. Survival of corneal allograft after treatment with anti-VLA-4 and anti-LFA-1 monoclonal antibodies

Treatment	Survival (weeks × n)
None	2 × 9, 3 × 2
Control mAb	2 × 5
Anti-VLA-4 mAb	3 × 3, 4 × 4
Anti-LFA-1 mAb	8, 9, 10, 12, 14, >15
Anti-VLA-4 + anti-LFA-1 mAbs	9 × 2, 11, >15 × 9*

Recipient mice were administered 0.5 mg of mAbs on days −2, 0, 1, 3, 5, 7 after surgery; *$p < 0.05$ vs other groups.

Table 2. Survival of primary skin allograft after treatment with anti-VCAM-1 and anti-VLA-4

Treatment	Survival (days)
None	14, 14, 14, 16, 17, 18
Control mAb	13, 14, 14, 15, 15, 17
Anti-VLA-4 + anti-VCAM-1 mAbs	13, 13, 15, 15, 16, 17

No significant prolongation of the primary skin graft was observed.

all allografts treated with anti-VLA-4 alone were rejected within 4 weeks, the survival rate treated with the two mAbs at 14 weeks was 75% and was significantly greater than that without treatment (Table 1). Cytotoxic responses to donor alloantigens were suppressed in mice treated with these two mAbs. Challenge test with second skin graft showed specific prolongation of donor strain skin, as compared to third-party strain skin, suggesting the specificity of this immunosuppression. This experiment also demonstrated the synergism between blockade of VLA-4- and LFA-1-dependent adhesion.

Effects of mAbs on VLA-4 and VCAM-1 were tested in primary skin allograft. However, we could not find any immunosuppressive effects, although various durations and doses of mAb treatment were tried (Table 2).

7 Unresponsiveness to Soluble Antigens

The same regimen of treatment was capable of inducing hyporesponsiveness to soluble antigens (Isobe et al. 1994). C3H/He mice were immunized with heat-aggregated human γ-globulin (HGG). Mice were injected with either saline, 100 μg of M18/2, PS/2, or M/K-2 or 50 μg each of both PS/2 and M/K-2 at the time of immunization. Booster immunization was performed 3 weeks after the initial immunization, and mice were bled at 24, 33, and 44 days after immunization. Mice injected with saline or control mAb produced antibodies to HGG, whereas antibody production was significantly suppressed in mice treated with PS/2 or a combination of both PS/2 and M/K-2. Treatment with M/K-2 alone showed intermediate results. Thus, mAbs to VLA-4 and VCAM-1 are capable of inducing immunological hyporesponsiveness not only to alloantigens but also to soluble antigens. The mechanism for this immunosuppression in B cell immunity remains to be investigated.

8 Regulation of Immunosuppression by Cytokines

The mechanism of the immunosuppression in cardiac allografts remains unclear. Since cytokines are a critical factor for immune regulation in transplantation, we analyzed cytokine profiles in mice that accepted cardiac allografts after treatment with anti-VCAM-1 and anti-VLA-4 mAbs. Expression of cytokine, which is produced by type 1 or 2 helper T (Th1, Th2) cells, was analyzed using in situ reverse transcriptase-polymerase chain reaction (RT-PCR) (NUOVO et al. 1994). Anti-VCAM-1 and anti-VLA-4 mAbs were administered for the first 5 days. Another group was treated with a 3 day course of anti-ICAM-1 plus anti-LFA-1 mAbs. Six control mice were treated daily with FK506 (0.1 mg/kg per day) and other mice received no treatment. The mRNA levels of Th1 cytokines (IFN-γ, IL-2) were enhanced in the cardiac allografts and spleens from nontreated mice. In the anti-ICAM-1/LFA-1 mAb treatment group, expression of Th2 (IL-4, IL-10) cytokines was significantly enhanced, resulting in the complete suppression of Th1 cytokine expression. In contrast, anti-VCAM-1/VLA-4 mAb treatment did not strongly influence the expression of Th2 cytokines (unpublished data). Both Th1 and Th2 expression were suppressed in grafts and spleens from the FK506-treated group.

Differential development of immature helper cells (Th0) to mature Th1 or Th2 cells is an important factor in determining the kinetics of cytokine production and immune responses to rejection (CHER and MOSSMAN 1987; FONG and MOSSMAN 1990). It has been shown that stimulation of Th0 cells in the absence of costimulation could anergize Th1 clones, while IL-4 production by Th2 is spared (GAJEWSKI et al. 1994). It is possible to assume that through such mechanisms antigen-specific Th2 clones are expanded while Th1 clones become inactivated (Fig. 7).

9 Effects on Chronic Rejection of Cardiac Allograft

Graft arteriopathy limits the long-term survival of allograft recipients (BOTAS et al. 1995; HOSENPUD et al. 1996). Murine cardiac allografts develop graft coronary arteriopathy similar to that observed in clinical chronic rejection. We have reported that treatment with short-term anti-ICAM-1 and anti-LFA-1 mAbs administration

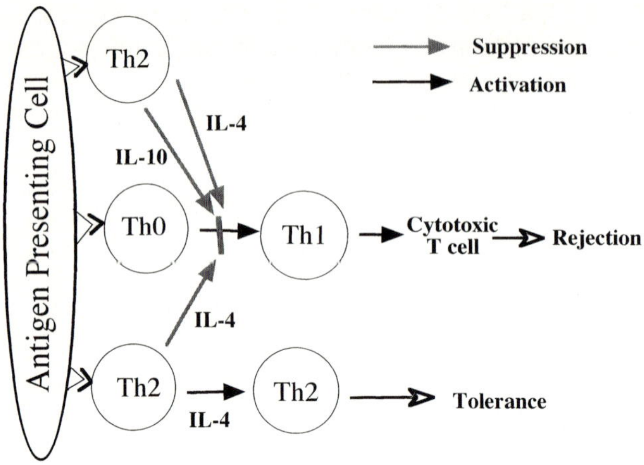

Fig. 7. Hypothesis on the association between activation of Th2 cytokines and tolerance induction

leads to specific tolerance of cardiac allografts as well as significant suppression of graft coronary arteriopathy in mice (SUZUKI et al., in press). We have observed that, in the experimental models, both VCAM-1 and ICAM-1 expression are enhanced in the thickened intima of graft coronary arteries (Fig. 8). We thus investigated the effects of VCAM-1/VLA-4 blockade on graft coronary arteriopathy.

Cardiac allografts were harvested at day 60 after treatment with anti-VCAM-1/VLA-4, anti-ICAM-1/LFA-1 or FK506. Anti-VCAM-1 plus anti-VLA-4 therapy resulted in a lower degree of intimal thickening than FK506 treatment with sup-

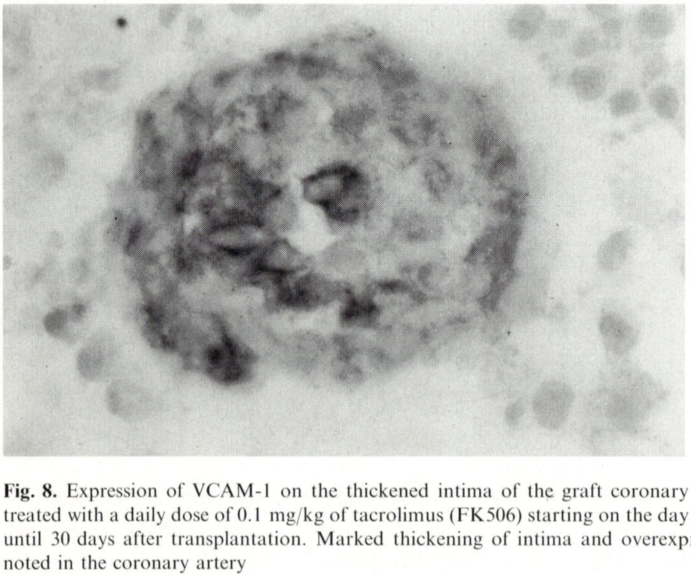

Fig. 8. Expression of VCAM-1 on the thickened intima of the graft coronary artery. The mouse was treated with a daily dose of 0.1 mg/kg of tacrolimus (FK506) starting on the day of heart transplantation until 30 days after transplantation. Marked thickening of intima and overexpression of VCAM-1 are noted in the coronary artery

pressed expression of ICAM-1, VCAM-1 and platelet-derived growth factor (PDGF)-B. However, no intimal thickening was observed in the grafts with anti-ICAM-1 plus anti-LFA-1 mAb treatment. These results indicate that blocking adhesion of VCAM-1 and VLA-4 prevents graft arteriopathy more effectively than FK506 treatment. However it does not induce the complete inhibition that anti-ICAM-1 and anti-LFA-1 provide. The inability to prevent graft arteriopathy by blocking VCAM-1 and VLA-4 adhesion may be due to incomplete suppression of growth factors and adhesion molecule expression.

10 Conclusion

Rejection is a complicated, poorly understood immunological process. The pathophysiological roles of VCAM-1/VLA-4 and ICAM-1/LFA-1 adhesion in acute as well as chronic rejection are still obscure. Although we demonstrated immunosuppression by blocking VCAM-1/VLA-4 adhesion, the majority of secondary skin grafts were rejected. These results indicate that immunological tolerance cannot be induced by this regimen, unlike ICAM-1/LFA-1 blockade. Knowledge of the preferential use of distinct costimulatory pathways in eliciting humoral and cellular immune responses to particular antigens may have important clinical implications. Our investigations, showing specific acceptance of allograft transplantation, have significant clinical implications for organ transplantation in the future.

Acknowledgement. The authors acknowledge Professors Morie Sekiguchi of Shinshu University, Ko Okumura of Juntendo University for their invaluable discussions. Monoclonal antibodies are kind gifts from Professor Ko Okumura. We would like to thank Misako Horii for her excellent technical assistance. This investigation was supported in part by grants-in-aid from The Ministry of Welfare and Health of Japan, The Ministry of Science, Culture, Sports and Education, Takeda Science Foundation, Bayer Cardiovascular Research Foundation, Japan Cardiovascular Research Foundation, and Yamanouchi Foundation for Research on Metabolic Disorders.

References

Allen MD, McDonald TO, Carlos T, Himes V, Fishbein D, Aziz S, Gordon D (1992) Endothelial adhesion molecules in heart transplantation. J Heart Lung Transplant 11:S8–13

Azuma M, Ito D, Yagita H, Okumura K, Phillips JH, Lanier LL, Somoza C (1993) B70 antigen is a second ligand for CTLA-4 and CD28. Nature 366:76–79

Botas J, Pinto FJ, Chenzbraun A, Liang D, Schroeder JS, Oesterle SN, Alderman EL, Popp RL, Yeung AC (1995) Influence of preexistent donor coronary artery disease on the progression of transplant vasculopathy. An intravascular ultrasound study. Circulation 92:1126–1132

Chavin KD, Qin L, Lin J, Kaplan AJ, Bromberg JS (1993) Anti-CD2 and anti-CD3 monoclonal antibodies synergize to prolong allograft survival with decreased side effects. Transplantation 55:901–908

Cher DJ, Mossman TR (1987) Two types of murine helper T cell clones. II. Delayed-type hypersensitivity is mediated by Th1 clones. J Immunol 138:3688–3694

Clevers H, Alarcon B, Wileman T, Terhorst C (1988) The T cell receptor/CD3 complex: a dynamic protein ensemble. Annu Rev Immunol 6:629

Cobbold SP, Jayasuriya A, Nash A, Prospero TD, Waldmann H (1989) Therapy with monoclonal antibodies by elimination of T-cell subsets in vivo. Nature 312:548–551

Cosimi AB, Conti D, Delmonico FL, Preffer FI, Wee S, Rothlein R, Faanes R, Colvin RB (1990) In vivo effects of monoclonal antibody to ICAM-1 (CD54) in nonhuman primates with renal allografts. J Immunol 144:4604–4612

Damle NK, Aruffo A (1991) Vascular cell adhesion molecule 1 induces T-cell antigen receptor-dependent activation of $CD4^+$ T lymphocytes. Proc Natl Acad Sci USA 88:6403–6407

Damle NK, Klussman K, Aruffo A (1992) ICAM-2, a second ligand for CD11a/CD18 (LFA-1), provides a costimulatory signal for T cell receptor-initiated activation of human T cells. J Immunol 148:665

Fong TAT, Mossman TR (1990) Alloreactive murine CD8+ T cell clone secrete the Th1 pattern of cytokines. J Immunol 144:1744–1752

Freeman GJ, Freedman AS, Segil JM, Lee G, Whitman JF, Nadler LM (1989) B7, a new member of neoplastic B cells. J Immunol 143:2714

Gajewski TF, Lancki DW, Stack R, Fitch FW (1994) "Anergy" of TH0 helper T lymphocytes induces downregulation of TH1 characteristics and a transition to a TH2-like phenotype. J Exp Med 179: 481–491

Harding FA, McAurthur JG, Gross JA, Raulet DH, Allison JP (1992) CD28-mediated signalling co-stimulates murine T cells and prevents induction of anergy in T-cell clones. Nature 356:607–609

Hession C, Moy P, Tizard R, Chisholm P, Williams C, Wysk M, Burkly L, Miyake K, Kincade P, Lobb R (1992) Cloning of murine and rat vascular cell adhesion molecule-1. Biochem Biophys Res Commun 183:163–169

Hori J, Isobe M, Yamagami S, Mizuochi S, Tsuru T (1997) Specific immunosuppression of corneal allograft rejection by combination of anti-VLA-4 and anti-LFA-1 monoclonal antibodies in mice. Exp Eye Res 65:89–98

Hosenpud JD, Novick RJ, Bennett LE, Keck BM, Fiol B, Daily OP (1996) The Registry of the International Society for Heart and Lung Transplantation: Thirteenth official report – 1996. J Heart Lung Transplant 15:655–674

Isobe M, Haber E, Khaw BA (1991) Early detection of rejection and assessment of cyclosporine therapy by indium-111 antimyosin imaging in mouse heart allografts. Circulation 84:1246–1255

Isobe M, Ihara A (1993) Tolerance induction against cardiac allograft by anti-ICAM-1 and anti-LFA-1 treatment: T cells respond to in vitro allostimulation. Transplant Proc 25:1079–1080

Isobe M, Narula J, Strauss HW, Khaw BA, Haber E (1992a) Imaging the rejecting heart: in vivo detection of MHC class II antigen induction. Circulation 85:738–746

Isobe M, Suzuki J, Yagita H, Okumura K, Yamazaki S, Nagai R, Yazaki Y, Sekiguchi M (1994) Immunosuppression to cardiac allografts and soluble antigens by anti-vascular cellular adhesion molecule-1 and anti-very late antigen-4 monoclonal antibodies. J Immunol 153:5810–5818

Isobe M, Suzuki J, Yamazaki S, Sekiguchi M (1996) Acceptance of primary skin graft after treatment with anti-ICAM-1 and anti-LFA-1 monoclonal antibodies in mice. Transplantation 62:411–413

Isobe M, Yagita H, Okumura K, Ihara A (1992b) Specific acceptance of cardiac allograft after treatment with anti-ICAM-1 and anti-LFA-1. Science 255:1125–1127

Issekutz TB (1991) Inhibition of in vivo lymphocyte migration to inflammation and homing to lymphoid tissues by the TA-2 monoclonal antibody. A likely role for VLA-4 in vivo. J Immunol 147:4178–4184

Jenkins MK, Taylor PS, Norton SD, Urdahl KB (1991) CD28 delivers a costimulatory signal involved in antigen-specific IL-2 production by human T cells. J Immunol 147:2461–2466

Kato K, Koyanagi M, Okada H, Takanashi T, Wong Y, Williams AF, Okumura K, Yagita H (1992) CD48 is a counter-receptor for mouse CD2 and is involved in T cell activation. J Exp Med 176: 1241–1249

Koulova L, Clark EA, Shu G, Dupont B (1991) The CD28 ligand B7/BB1 provides costimulatory signal for alloactivation of $CD4^+$ T cells. J Exp Med 173:759

Larsen CP, Elwood ET, Alexander DZ, Ritchie SC, Hendrix R, Tucker-Burden C, Cho HR, Aruffo A, Hollenbaugh D, Linsley PS, Winn KJ, Pearson TC (1996) Long-term acceptance of skin and cardiac allografts after blocking CD40 and CD28 pathways. Nature 381:434–438

Lenschow DJ, Zeng Y, Thistlethwaite JR, Montag A, Brady W, Gibson MG, Linsley PS, Bluestone JA (1992) Long-term survival of xenogeneic pancreatic islet grafts induced by CTLA4Ig. Science 257:789–792

Linsley PS, Brady W, Grossmaire L, Aruffo A, Damle NK, Ledbetter JA (1991) Binding of the B cell activation antigen B7 to CD28 costimulates T-cell proliferation and interleukin-2 mRNA accumulation. J Exp Med 173:721

Miyake K, Medina K, Ishihara K, Kimoto M, Auerbach R, Kincade PW (1991a) A VCAM-like adhesion molecule on murine bone marrow stromal cells mediates binding of lymphocyte precursors in culture. J Cell Biol 114:557–565

Miyake K, Weissman IL, Greenburger JS, Kincade PW (1991b) Evidence for a role of the integrin VLA-4 in lympho-hemopoiesis. J Exp Med 173:599–607

Moingeon PE, Lucich JL, Stebbins CC, Recny MA, Wallner BP, Koyasu S, Reinherz EL (1991) Complementary roles for CD2 and LFA-1 adhesion pathways during T cell activation. Eur J Immunol 21:605–610

Mueller D, Jenkins M, Schwartz R (1989) Clonal expansion versus functional clonal inactivation: a costimulatory outcome of T cell antigen receptor occupancy. Annu Rev Immunol 7:445–480

Nakakura EK, McCabe SM, Zheng B, Shorthouse RA, Scheiner TM, Blank G, Jardieu PM, Morris RE (1993) Potent and effective prolongation by anti-LFA-1 monoclonal antibody monotherapy of non-primarily vascularized heart allograft survival in mice without T cell depletion. Transplantation 55:412–417

Nicolls MR, Aversa GG, Pearce NW, Spinelli A, Berger MF, Gurley KE, Hall BM (1993) Induction of long-term specific tolerance to allografts in rats by therapy with an anti-CD3-like monoclonal antibody. Transplantation 55:459–468

Nuovo GJ, Gallery F, MacConnell P, Braun A (1994) In situ detection of polymerase chain reaction-amplified HIV-1 nucleic acids and tumor necrosis factor-a RNA in the central nervous system. Am J Pathol 144:659–666

Ono K, Lindsey ES (1969) Improved technique of heart transplantation in rats. J Thorac Cardiovasc Surg 57:225–229

Orosz CG, Ohye RG, Pelletier RP, Buskirk AM, Haung E, Morgan C, Kincade PW, Ferguson RM (1993) Treatment with anti-vascular cell adhesion molecule 1 monoclonal antibody induces long-term murine cardiac allograft acceptance. Transplantation 56:453–460

Osborn L, Hession C, Tizard R, Vassallo C, Luhowskyj S, Chirosso G, Lobb R (1989) Direct expression cloning of vascular cell adhesion molecule-1, a cytokine-induced endothelial protein that binds to lymphocytes. Cell 59:1203–1211

Pelletier R, Ohye RP, Kincade P, Ferguson R, Orosz C (1993) Monoclonal antibody to anti-VCAM-1 interferes with murine cardiac allograft rejection. Transplant Proc 25:839–841

Qin S, Cobbold SP, Benjamin R, Waldmann H (1989) Induction of classical transplantation tolerance. J Exp Med 169:779–794

Sanchez MF, Simon P, Thompson S, Springer TA (1983) Mapping of antigenic and functional epitopes on the alpha- and beta-subunits of two related mouse glycoproteins involved in cell interactions, LFA-1 and Mac-1. J Exp Med 158:586–602

Schwartz RH (1990) A cell culture model for T lymphocyte clonal anergy. Science 248:1349–1356

Schwartz RH, Mueller DL, Jenkins MK, Quill H (1989) T-cell clonal anergy. Cold Spring Harb Symp Quant Biol 2:605–610

Shimizu Y, van Seventer GA, Horgan KJ, Shaw S (1990a) Costimulation of proliferative responses of resting $CD4^+$ T cells by the interaction of VLA-4 and VLA-5 with fibronectin or VLA-6 with laminin. J Immunol 145:59–67

Shimizu Y, van Seventer GA, Horgan KJ, Shaw S (1990b) Roles of adhesion molecules in T-cell recognition: fundamental similarities between four integrins on resting human T cells (LFA-1, VLA-4, VLA-5, VLA-6) in expression, binding, and costimulation. Immunol Rev 114:109–143

Simmons D, Makgoba MW, Seed B (1989) An LFA-3 cDNA encodes a phospholipid-linked membrane protein homologous to its receptor CD2. Nature 329:840

Springer TA (1990) Adhesion receptors of the immune system. Nature 346:425–433

Springer TA, Dustin ML, Kishimoto TK, Marlin SD (1987) The lymphocyte function-associated LFA-1, CD2, and LFA-3 molecules: cell adhesion receptors of the immune system. Annu Rev Immunol 5:223–252

Steinhoff G, Behrend M, Schrader B, Duijvestijn AM, Wonigeit K (1993) Expression patterns of leukocyte adhesion ligand molecules on human liver endothelia. Lack of ELAM-1 and CD62 inducibility on sinusoidal endothelia and distinct distribution of VCAM-1, ICAM-1, ICAM-2, and LFA-3. Am J Pathol 142:481–488

Suzuki J, Isobe M, Yamazaki S, Horie S, Okubo Y, Sekiguchi M (1997) Short-term blockade of ICAM-1 and LFA-1 prevents cardiac allograft arteriopathy in mice. J Heart Lung Transplant (in press)

Tanio JW, Basu CB, Albelda SM, Eisen HJ (1994) Differential expression of the cell adhesion molecules ICAM-1, VCAM-1, and E-selectin in normal and posttransplantation myocardium. Cell adhesion molecule expression in human cardiac allografts. Circulation 89:1760–1768

Taylor PM, Rose ML, Yacoub MH, Pigott R (1992) Induction of vascular adhesion molecules during rejection of human cardiac allografts. Transplantation 54:451–457

van Seventer GA, Shimizu Y, Horgan KJ, Shaw S (1990) The LFA-1 ligand ICAM-1 provides an important costimulatory signal for T cell receptor-mediated activation of resting T cells. J Immunol 144:4579–4586

van Seventer GA, Newman W, Shimizu Y, Nutman TB, Tanaka Y, Horgan KJ, Gopal TV, Ennis E, O'Sullivan D, Grey H, Shaw S (1991) Analysis of T cell stimulation by superantigen plus major histocompatibility complex class II molecules or by CD3 monoclonal antibody: costimulation by purified adhesion ligands VCAM-1, ICAM-1, but not ELAM-1. J Exp Med 174:901–913

Weaver CT, Unaue ER (1990) The costimulatory function of antigen-presenting cells. Immunol Today 11:49

Wood MJ, Sloan DJ, Dallman MJ, Charlton HM (1993) Specific tolerance to neural allografts induced with an antibody to the interleukin 2 receptor. J Exp Med 177:597–603

Virus-Induced T Cell Activation and the Inflammatory Response

A.R. Thomsen, A. Nansen, and J.P. Christensen

1 Introduction 99
2 The Experimental Model 100
3 The Virus-Specific T Cell Response 101
4 Bystander T Cell Activation 102
5 Phenotypic Characterization of Virus-Activated T Cells 104
6 Functional Characterization of Virus-Activated T Cells 106
7 Consequences for T Cell Migration and the Inflammatory Response 109
8 Role of Cytokines 114
9 An Integrated Model for Effector T Cell Homing to Sites of Viral Replication 116
10 Implications 117
References 119

1 Introduction

Effector T cells constitute a critical component in the immune response to most viral infections (Doherty et al. 1992). Consequently, understanding the mechanisms regulating the generation and function of effector T cells is central to the study of viral pathogenesis. It is characteristic that T cells – as opposed to B cells – are only able to exert their effector function within a very limited distance. Therefore, once effector T cells are generated, they must be able to migrate to relevant sites of infection. This requires a set of surface receptors which direct the migration of effector cells to infected areas. However, not only must fully differentiated effector T cells be able to reach any part of the organism, but, it is also nescessary for naive T cells to continually recirculate in order for the immune system to optimally utilize the limited number of cells with T cell receptors (TCRs) relevant to a given antigen.

Institute of Medical Microbiology and Immunology, University of Copenhagen, The Panum Institute, 3C Blegdamsvej, 2200 Copenhagen, Denmark

Of course, the recirculation pattern of naive cells needs to be different from that of effector cells since the triggering of the naive T cell is a carefully controlled event confined to specialized tissues – the secondary lymphoid organs – containing high numbers of antigen-presenting cells (APCs). Indeed, the interaction of naive cells with foreign antigen outside of these organs might be dangerous since antigen-recognition in the absence of a costimulatory signal will lead to anergy or deletion (Matzinger 1994). Thus, the continual passage of naive cells through the lymph nodes and spleen serves to allow all relevant T cells to come into contact with a foreign antigen under conditions which are optimal for triggering of an effective immune response (Kündig et al. 1995).

This ongoing surveillance of the draining lymph nodes presupposes the expression of specialized surface receptors which are superfluous on fully differentiated effector T cells. Consequently, the phenotypic changes seen in association with T cell activation include fundamental changes in the expression of adhesion molecules involved in regulating lymphocyte circulation. Thus expression of L-selectin, known to be mandatory for homing to the lymph nodes through the high endothelial venules (Bradley et al. 1994; Xu et al. 1996; Arbones et al. 1994), decreases as a result of T cell activation (Jung et al. 1988; Mobley and Dailey 1992; Bradley et al. 1991). At the same time expression of the integrins LFA-1 and VLA-4 increases (Mobley and Dailey 1992; Issekutz 1991), consistent with the dogma that these molecules are important for extravasation at sites of inflammation (Issekutz 1992). However, for specific targeting of effector cells to infected areas, local changes in and around the transversing vessels are also required (Butcher 1991; Springer 1994), such as up-regulation of appropriate ligands on the endothelium as well as production of various cytokines (Issekutz 1990).

In this review we describe the changes in expression of adhesion molecules that are induced in the context of an immune response to a systemic viral infection. Further, since a changed expression of adhesion molecules not only signals an altered migration pattern, but marks a more general change in the functional status of the lymphocyte (Mackay 1991; Butcher and Picker 1996), we will also take the opportunity to describe the profound perturbation of T cell function that is often associated with systemic viral infections, with emphasis on the consequences for the formation of the inflammatory exudate and T cell effector capacity.

2 The Experimental Model

Most of the data on which we base this review have been obtained in mice infected with the arenavirus lymphocytic choriomeningitis virus (LCMV) (Pfau and Thomsen 1993). LCMV is a noncytolytic virus that induces little or no inflammation in T cell deficient mice (Doherty and Zinkernagel 1974; Allan et al. 1987; Marker et al. 1995; Campbell et al. 1994). In immunocompetent mice, however, an inflammatory reaction is found in infected organs. This reaction is the result of a potent T cell response dominated by the generation of $CD8^+$ effector cells (Doherty et al. 1990; Leist et al. 1987). In turn these cells initiate the formation of an inflammatory exudate which in the acute phase consists of mostly $CD8^+$ T cells and mononuclear phagocytes (Ceredig et al. 1987). With time some $CD4^+$ T cells and B cells may also be attracted to sites of LCMV infection (Christensen et al. 1995; Moskophidis et al. 1990). Because infection with this virus is associated with little nonspecific inflammation, this is an ideal model for studying the mechanisms underlying the formation of a virus-induced, T cell dependent inflammatory exudate. Furthermore, since LCMV readily infects the meninges, and in intracerebrally (i.c.) infected mice induces a severe inflammatory reaction in the cerebrospinal fluid (CSF) (Allan et al. 1987; Andersen et al. 1990) that normally contains only very few leukocytes, we have here a perfect site for obtaining effector cells recruited to an infected area in the context of an antiviral T cell response (Carp et al. 1971; Doherty 1973; Ceredig et al. 1987). Such cells may then be further analyzed without introducing artifacts resulting from, e.g., the enzymatic disruption of tissue that is necessary when using solid organs.

3 The Virus-Specific T Cell Response

When mice are infected systemically with LCMV, substantial clonal expansion of virus-specific cytotoxic T lymphocyte precursors (CTLps) takes place. The frequency of these cells reaches a peak around day 8–10 postinfection, at which time about 1/30–1/100 spleen cells is an LCMV-specific CTLp (Lau et al. 1994). This expansion phase is followed by a decline in precursor frequency to about 1/1000 spleen cells, and the frequency then remains relatively stable at this level for what is probably the rest of life (Lau et al. 1994). In the course of the infection, part of the precursor cells undergo differentiation to effector CTLs. This CTL response, which can be measured directly ex vivo, also peaks around day 8–10 postinfection (Marker and Volkert 1973). Coinciding with the appearance of virus-specific CTLs in the lymphoid organs, organ virus titers start to decline and after about 4 weeks little or no virus can be detected in the animals (Marker and Volkert 1973; Thomsen and Marker 1989). Virus clearance as well as the immunopathology of this infection is temporally associated with the influx of $CD8^+$ CTLs into infected organs (e.g., liver or meninges) (McIntyre and Welsh 1986; Andersen et al. 1990; Allan et al. 1987; Leist et al. 1987; Moskophidis et al. 1987; Zinkernagel and Doherty 1973; Zinkernagel et al. 1986; Kagi et al. 1994) and in $CD8^+$ T cell deficient mice neither virus control nor a substantial inflammatory response is observed (Christensen et al. 1994; Doherty et al. 1993; Lehmann-Grube et al. 1993). Thus the migration of effector T cells from sites of production to infected organs is a central event which determines the outcome of this infection. For example, it has been demonstrated that splenectomy will delay death in i.c. infected mice (Doherty and Zinkernagel 1974), which otherwise die as a result of the immune attack on infected structures in brain.

4 Bystander T Cell Activation

In addition to the generation of virus-specific effector T cells, LCMV infection is associated with marked T cell proliferation leading to the enlargement of lymphoid organs and lymphocytosis (Fig. 1). Most of the proliferating cells are $CD8^+$ T cells (Kasaian and Biron 1989; Christensen et al. 1996b), many of which transiently express the high affinity form of the interleukin-2 receptor (IL-2R; CD25/CD122) and proliferate in response to low doses of IL-2 in vitro (Andersson et al. 1995; Lynch et al. 1989). Since at the peak of the response up to one third of splenic $CD8^+$ cells may be found to have entered the S or G2+M phase (Fig. 2), it is evident from comparison with the above CTLp frequencies that the majority of these cells are very unlikely to be LCMV-specific T cells, even if one allows a wide margin for in vitro seeding efficiency of precursor cells in the acute phase of the infection (at this stage many LCMV-specific T cells may be terminally differentiated, and thus unable to undergo further clonal expansion; Ceredig et al. 1987). As fundamentally similar results have been obtained in several mouse strains as well as in mice infected with other, very different viruses (Pichinde, vesicular stomatitis virus, influenza), indications are that nonspecific (bystander) activation of $CD8^+$ T cells is a general phenomenon occurring in the context of most viral infections (Doherty et al. 1994).

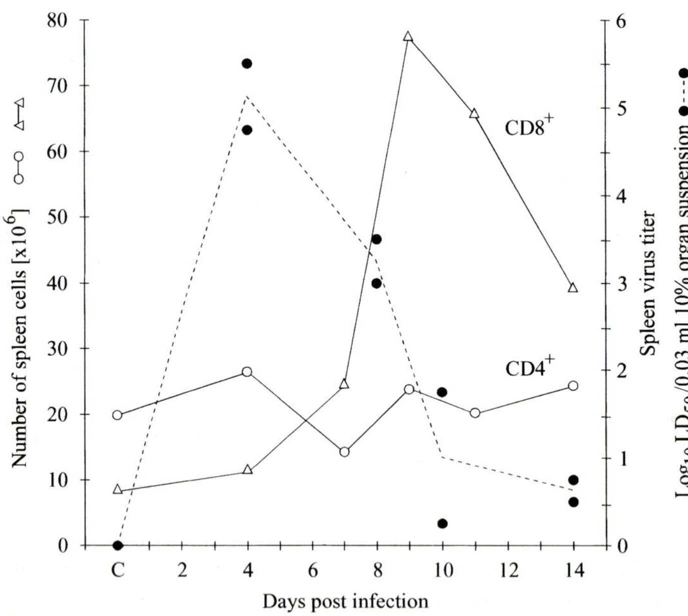

Fig. 1. Numbers of splenic $CD4^+$ and $CD8^+$ T cells as a function of time after infection with lymphocytic choriomeningitis virus (LCMV); for comparison, the time course of splenic virus titers are also presented. (From Christensen et al. 1996b)

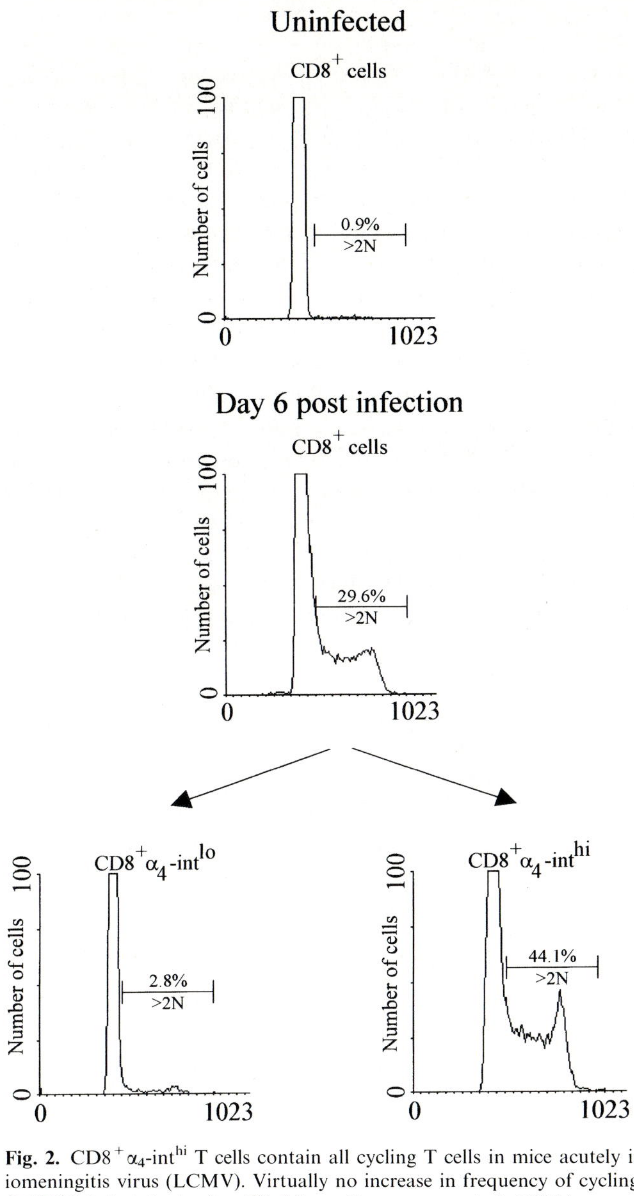

Fig. 2. CD8$^+$ α_4-int^{hi} T cells contain all cycling T cells in mice acutely infected with lymphocytic choriomeningitis virus (LCMV). Virtually no increase in frequency of cycling CD4$^+$ T cells are observed in (LCMV)-infected mice (modified from Christensen et al. 1996b)

The mechanism underlying this phenomenon of virus-induced bystander T cell activation is a much debated subject, pertinent as to how T cell memory is maintained (Doherty et al. 1994; Beverley 1990; Tough et al. 1996; Selin et al. 1994; Ahmed and Gray 1996). Viral superantigens could be thought to be involved, but at least in LCMV-infected mice no Vβ preference is observed (Andersson et al.

1995; NAHILL and WELSH 1993), rendering this unlikely as a general explanation. Alternatively, polyclonal T cell activation could be the result of cross-reactive, low affinity interactions involving the clonotypic TCRs (BEVERLEY 1990). If such low affinity interactions were important, one would expect to find that the expression of adhesion molecules involved in cell-cell interaction (e.g., LFA-1) was a limiting factor, thus favoring the activation of memory T cells which express these molecules at a higher level than do naive cells (SPRINGER 1990). Consistent with this prediction, it has been found that memory cells to unrelated viruses may be activated during an antiviral immune response (YANG et al. 1989; NAHILL and WELSH 1993; SELIN et al. 1994). However, in studies involving mutant mice with hypomorphic expression of CD18 (β_2-integrin) as well as ICAM-1 deficient mice, we did not observe a reduced polyclonal response (CHRISTENSEN et al. 1996a). This finding seems to argue against a predominant role of low affinity interactions as well as cognate cell-cell interactions mediated through these adhesion molecules (BROD et al. 1990; VYTH-DREESE et al. 1993). Nonspecific stimulation resulting from the cytokines released in the context of viral infections thus appears to be the preferred alternative when trying to explain most of the bystander activation. TOUGH et al. (1996) have suggested that virus-induced interferon (IFN)-α/β might be critical, based on the finding that memory $CD8^+$ T cells start to proliferate in response to injection of poly I:C, an artificial homologue of viral RNA and a potent inducer of IFN-α/β. However, in contrast to genuine virus-induced T cell activation, this regimen does not induce the expression of CD25 (IL-2R α-chain) on activated T cells, suggesting that IFN-α/β constitutes at most one of the signals underlying virus-induced polyclonal T cell activation. In contrast, much reduced T cell proliferation is observed in IL-2 deficient mice, indicating a critical role for this cytokine (COUSENS et al. 1995). In this context it is notable that IL-2 in vitro has been found to induce T cells to undergo essentially similar phenotypic changes as observed in vivo in virus-infected mice (ROTH 1994; ANDERSSON et al. 1994). The source of IL-2 remains obscure; at least $CD4^+$ T cells are not pivotal because deficiency of $CD4^+$ cells as seen in antibody-depleted or class II-deficient mice does not substantially reduce bystander activation (ANDERSSON et al. 1995; CHRISTENSEN et al. 1994; KASAIAN et al. 1991). In contrast to IL-2, IFN-γ is not limiting, as IFN-γ deficient mice may respond at least as well as wild-type mice (unpublished data).

5 Phenotypic Characterization of Virus-Activated T Cells

The result of the pronounced virus-induced $CD8^+$ T cell expansion is the generation of a distinct subset of T cells with a phenotypic profile characteristic of activated T cells (Fig. 2 and 3). Thus, whereas phenotypically naive (i.e., α_4-int^{lo} LFA-1^{lo}Pgp-1^{lo}L-sel^{hi}) T cells dominate in young, uninfected mice, 65%–85% of the $CD8^+$ T cells present in the spleen and lymph nodes of mice infected i.v. with LCMV 8–10 days earlier are α_4-int^{hi}LFA-1^{hi}Pgp-1^{hi}L-sel^{lo} (ANDERSSON et al.

1994), and many of these cells transiently express other markers of activation besides high affinity IL-2R (CD25/CD122), reduced levels of CD45RB and increased expression of CD71 and Mac-1, the expression of which seems to be associated with recent activation (ANDERSSON et al. 1994, 1995; MCFARLAND et al. 1992). In addition, analysis of low-angle and side scatter patterns revealed an increase in average cell size and granularity of these cells.

Following virus control, marked contraction of the $CD8^+$ T cell pool takes place, and many of the primed cells undergo apoptosis probably as a result of activation-induced cell death (CHRISTENSEN et al. 1996b; RAZVI et al. 1995a). However, a significant number of virus-activated cells are found in the spleen and blood for several months. At this stage most of the cells have returned to the size of small lymphocytes, but are still characterized by a primed phenotype: α_4-int^{hi}LFA-1^{hi}Pgp-1^{hi} (α_4-integrin expression is three to four times lower than on recently activated cells, but clearly higher, i.e., three to five times, than on naive cells; ANDERSSON et al. 1995).

The majority of these cells expresses low levels of L-selectin, but we also find a small subpopulation of T cells expressing high levels of α_4-integrin and LFA-1 together with moderate to high levels of L-selectin (Fig. 3). Based on the recent finding that primed T cells may regain L-selectin with time from antigenic stimulation (TRIPP et al. 1995; MOBLEY et al. 1994; RAZVI et al. 1995b), we interpret this subset to represent cells in this phase. With time all phenotypic evidence of previous exposure to viral challenge tends to disappear. The precise kinetics of disappearance depends on the virus infection analyzed. Thus, we have compared the pattern found in mice infected with two strains of LCMV, differing in their ability to persist at low levels in immunized mice, to mice infected with VSV, which causes an acute infection characterized by limited viral replication. This analysis showed that while α_4-int^{hi}L-sel^{lo} cells are induced in all cases, albeit to a lower degree in VSV-infected mice, hardly any cells maintained that phenotype for long in the latter mice. This is in striking contrast to LCMV-infected mice in which primed cells may be found for at least up to 3 months after infection, with a tendency for even more prolonged persistence in mice infected with the strain being most difficult to eliminate (unpublished data). Since the majority of the virus-induced cells cannot be virus-specific (note that $CD8^+$ α_4-int^{hi} cells constitute about 10%–15% of splenic cells at 2 months postinfection whereas the frequency of LCMV-specific CTLps is about 1/1000 at this time), these findings indicate that bystander activation induces a large number of T cells to follow the same pattern of phenotypic changes believed to occur for specifically primed cells. Indeed recent analysis of virus-specific TCR transgenic cells demonstrate that these cells undergo exactly the same pattern following stimulation with the specific antigen (ZIMMERMAN et al. 1996). Furthermore, the persistence of cells with a primed phenotype appear to correlate with the persistence of antigen, strongly suggesting that maintenance of this subset requires some kind of ongoing stimulation. Consistent with this, we have found that cells belonging to this subset are more susceptible to treatment with the cell-cycle specific drug hydroxyurea than are cells with a naive phenotype, indicating a higher level of cell cycling (CHRISTENSEN et al. 1996b).

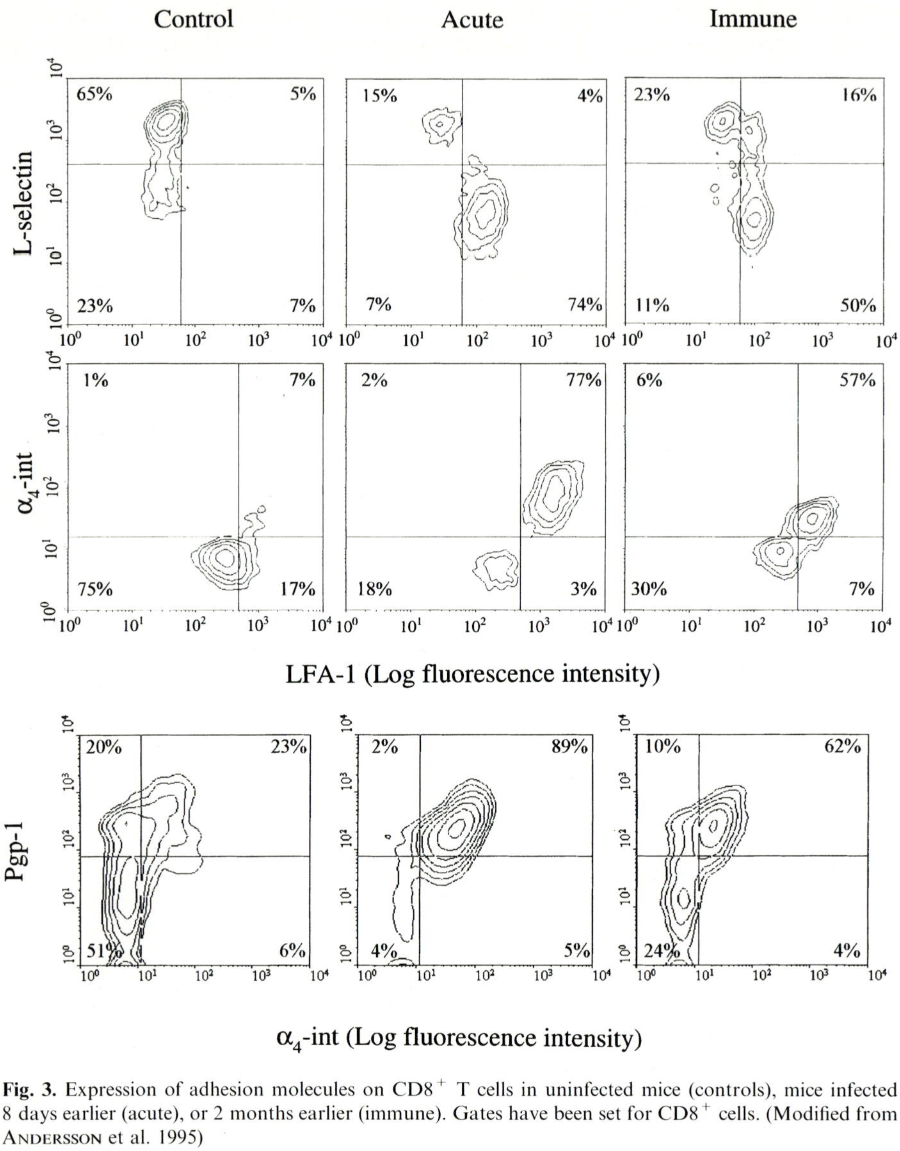

Fig. 3. Expression of adhesion molecules on CD8$^+$ T cells in uninfected mice (controls), mice infected 8 days earlier (acute), or 2 months earlier (immune). Gates have been set for CD8$^+$ cells. (Modified from ANDERSSON et al. 1995)

6 Functional Characterization of Virus-Activated T Cells

In order to clarify the functional status of the virus-induced T cell subset delineated above, spleen cells from LCMV-infected mice were sorted according to expression of relevant adhesion molecules and their ability to exert cell lysis and to produce IFN-γ was evaluated (summarized in Fig. 4). As expected, all *LCMV-specific* CTLs from acutely infected mice were found to be α_4-int^{hi}LFA-1^{hi}L-sel^{lo} (ANDERSSON

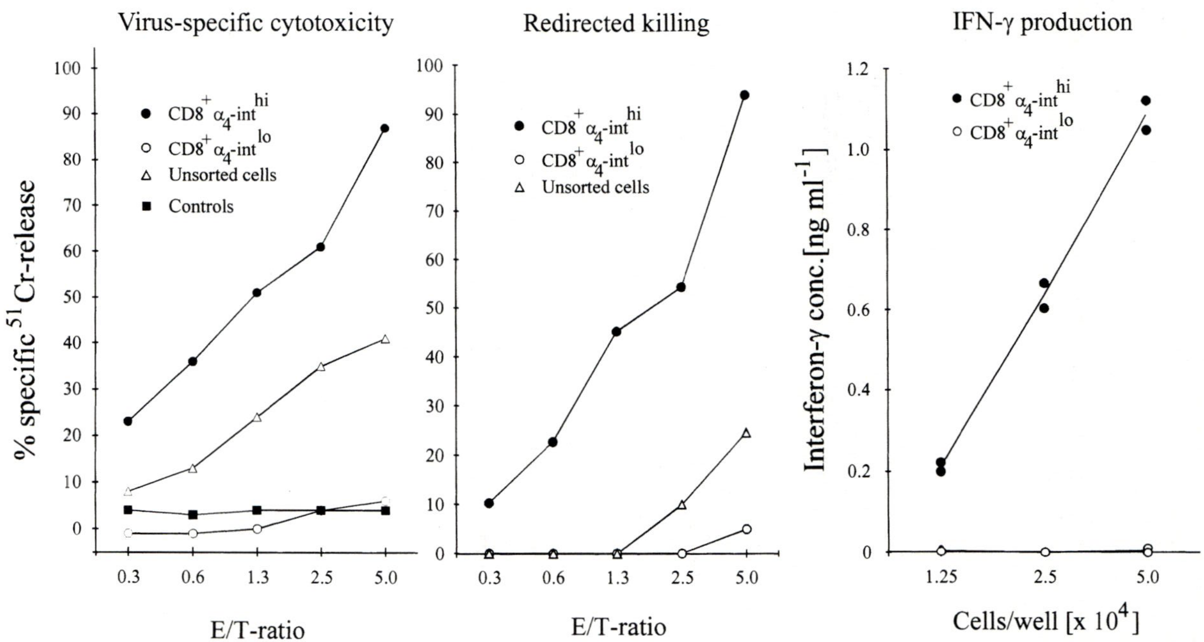

Fig. 4. Functional capacity of CD8$^+$ α_4-int^{hi} T cells in mice acutely infected with lymphocytic choriomeningitis virus (LCMV). CD8$^+$ T cells from mice infected 7 days earlier were sorted into α_4-int^{hi} and α_4-int^{lo} cells and tested for virus-specific cytotoxicity, redirected killing, and capacity to produce interferon (IFN)-γ following short-term stimulation with anti-CD3. (Modified from ANDERSSON et al. 1994 Copyright 1994, the American Association of Immunologists; CHRISTENSEN et al. 1996c)

et al. 1994, 1995; RAZVI et al. 1995b). Matching results were obtained in other systems involving viral as well as alloantigenic stimulation (HOU and DOHERTY 1993; MOBLEY et al. 1994). Consistent with this, TCR transgenic cells have been found to acquire the above phenotype when stimulated in vivo with the relevant antigen (ZIMMERMAN et al. 1996).

Taken together, these findings strengthen the assumption that this phenotype signifies activated/primed T cells. However, due to the fact that their numbers in virus-infected animals much exceed what can reasonably be virus-specific CTLs, further analysis was carried out. In previous studies it had been found that viral infections often induce a polyclonal CTL response involving alloreactive cells and memory cells with specificity for unrelated viruses (YANG et al. 1989). Moreover, $>50\%$ of $CD8^+$ T cells in the spleen of LCMV-infected mice contain granules that stain for esterase activity and perforin (KRAMER et al. 1989). Given the scatter pattern of the activated T cell subset these are likely to represent overlapping or identical cell subsets. Therefore, sorted cells were also analyzed in a redirected killing assay that detects *all* activated T cells with cytotoxic capacity. Again it was found that these cells were α_4-int^{hi}LFA-1^{hi} (expression of L-selectin was not evaluated) (CHRISTENSEN et al. 1996c). In addition, a comparison of splenocytes from mice infected with LCMV or VSV revealed that the percentage of activated cells as revealed by flowcytometry (LCMV > VSV) correlated with activity in the redirected killing assay (LCMV > VSV) (unpublished data), suggesting that many of the phenotypically activated cells also had aquired effector cell status. However, as a precise evaluation would require analysis of single cells, we exploited the fact that LCMV infection is associated with a highly polarized type I cytokine profile and that many T cells have been found at the mRNA level to be primed for production of IFN-γ (COLLE et al. 1993). Sorting analysis demonstrated that most cells producing IFN-γ in response to short-term stimulation with anti-CD3 in vitro were $CD8^+$ α_4-int^{hi}L-sel^{lo} cells (CHRISTENSEN et al. 1996c). Staining of cells from parallel cultures for presence of intracellular cytokine revealed that about half of splenic $CD8^+$ α_4–int^{hi} cells from acutely infected mice were positive for IFN-γ (Fig. 5), substantiating that many of these cells have indeed differentiated towards effector cell status. That many of the activated cells are terminally differentiated effectors is also suggested by the observation that anti-CD3 induced proliferation is inversely related to anti-CD3 stimulated IFN-γ production (CHRISTENSEN et al. 1996c). Thus, it may be concluded that virus-induced bystander activation not only induces blast transformation and cycling of $CD8^+$ T cells, but also drives many of the generated cells to differentiate into effector cells. This is of considerable interest since the signals inducing these different steps appear to be different, with effector cell differentiation being most demanding (BACHMANN et al. 1996; DOHERTY et al. 1994).

Based on the fact that T cells with the phenotype of primed cells persist for several months in LCMV-infected mice (Fig. 3), it was also of interest to establish the functional status of these cells late in the infection. An augmented capacity to excert cytolysis (albeit at a much lower level than during the acute phase of the infection) may be found for several months following LCMV challenge, and the same has recently been found to hold true for production of IFN-γ (MARKER and

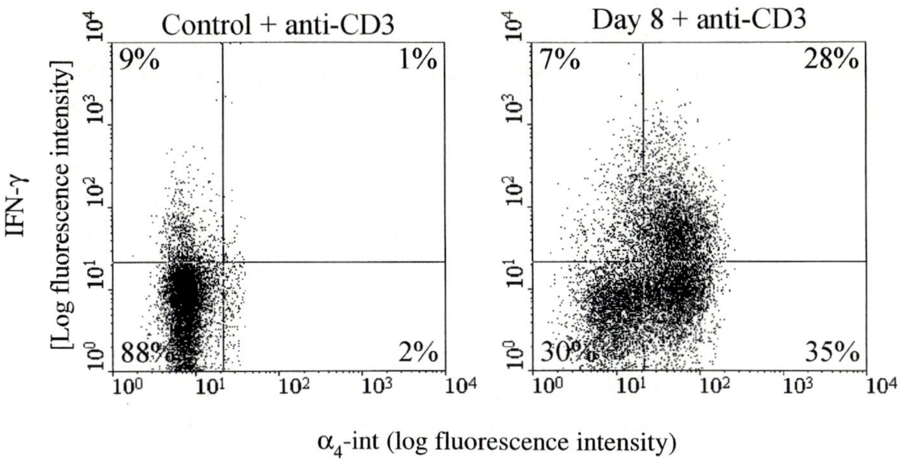

Fig. 5. Expression of interferon (IFN)-γ in CD8$^+$ T cells in mice acutely infected with lymphocytic choriomeningitis virus (LCMV). Spleen cells were stimulated in vitro with anti-CD3 for 6 h, stained for expression of α_4-integrin and CD8, permeabilized and stained for IFN-γ. Gates have been set for CD8$^+$ cells. (From CHRISTENSEN et al. 1996c)

VOLKERT 1973; CHRISTENSEN et al. 1996c). This is consistent with earlier indications that ongoing immune surveillance is needed to permanently control this infection (VOLKERT and LUNDSTEDT 1968). Sorting of splenic T cells into CD4$^+$ and CD8$^+$ cells revealed that only CD8$^+$ cells from mice infected 2 months earlier had a substantially increased capacity to produce IFN-γ compared to T cells from uninfected controls. Separation of the CD8$^+$ subset into Pgp-1^{hi} and Pgp-1^{lo} cells further revealed that the former subset was the only one to contain primed cells, as evaluated both for production of IFN-γ and for killing activity in the redirected assay (CHRISTENSEN et al. 1996c). Since it was previously found that this subset contains all LCMV-specific memory CTLps (TABI et al. 1987; LAU et al. 1994), these results indicate that the long-standing increase in cells with a primed phenotype represents the accumulation of previously activated cells, some of which still appear to be in a "poised" state. In this context it is of note that the frequency of activated/primed cells in virus-primed (LCMV and VSV) mice, as evaluated by flow cytometry, correlates with cytolytic activity measured by redirected killing as well as with the capacity to rapidly respond to antigenic challenge with T cell dependent inflammation (unpublished data; KÜNDIG et al. 1992).

7 Consequences for T Cell Migration and the Inflammatory Response

The extravasation of leukocytes at sites of inflammation is generally viewed as a stepwise process involving several receptor/ligand interactions (BUTCHER 1991;

Springer 1994; Butcher and Picker 1996). The first step appears to be a loose interaction with the endothelium resulting in the rolling of the leukocyte along the interior surface of the vessel. Molecules of the carbohydrate binding selectin family are believed to play a major role at this stage. In the next step the leukocyte is thought to be triggered by engagement of appropriate receptors which leads the leukocyte to markedly change its adhesive properties. Integrins, particularly LFA-1 and VLA-4, are important at this stage, mediating strong adhesion to the endothelium. The signals regulating integrin binding affinity are not clearly established, but various chemokines and their receptors are considered to be important (Mackay 1996). Finally, the leukocyte migrate through opposing endothelial cells into the surrounding tissue.

Although there are still many uncertainties regarding details of lymphocyte extravasation (Shimizu et al. 1992; Butcher and Picker 1996), this basic model provides us with sufficient insight to appreciate the potential significance of the changes in expression of adhesion molecules that is induced on T cells in the context of a viral infection. Thus, cells with high expression of VLA-4 and LFA-1 would clearly be suited for homing to inflamed areas, and even phagocytic glycoprotein-1 (Pgp-1) may be of relevance in this context (Camp et al. 1993). Furthermore, the down-regulation of L-selectin will prevent the waste of important effector cells associated with their passage through uninvolved lymph nodes (lymph nodes draining infected areas would still be reached by effector cells via the afferent lymphatics and, in addition, would probably in themselves represent inflamed areas; Mackay et al. 1992). A number of experimental observations obtained in the course of our studies on the murine LCMV infection fit this interpretation.

First, T cells isolated from virus-infected, inflamed tissues, e.g., in casu CSF cells from i.c. infected mice (Fig. 6) or peritoneal exudate cells from i.p. infected animals, are exclusively α_4-int^{hi}LFA-1^{hi}Pgp-1^{hi} (Andersson et al. 1994, 1995; Ceredig et al. 1987; Christensen et al. 1996c). Second, the influx of T cells correlate temporally with generation of α_4-int^{hi}LFA-1^{hi} cells in the spleen (Andersson et al. 1994; Christensen et al. 1995). Third, and perhaps most convincing, the marked bias for CD8$^+$ T cells noted in virus-infected organs matches the striking difference in phenotypic pattern observed for CD4$^+$ and CD8$^+$ T cells in the lymphoid organs of virus-infected animals (Andersson et al. 1994; Christensen et al. 1995). Thus, while CD8$^+$ T cell activation leads to generation of α_4-int^{hi}LFA-1^{hi} effector cells, few CD4$^+$ T with this phenotype are generated. Notably, the bias for CD8$^+$ T cells does not merely reflect their numerical dominance, as this bias is observed in infected organs before it occurs in the spleen, but clearly correlates with the predominance of CD8$^+$ cells amongst splenic α_4-int^{hi} T cells (Christensen et al. 1995). In addition to these studies carried out in T cell high responder mice, we also analyzed the inflammatory exudate in low responder mice, which undergo a more chronic infection. In this case even CD4$^+$ T cells and B cells are found in the inflammatory exudate, and again the inflammatory cells are found to be recruited predominantly from the α_4-int^{hi} subset (Christensen et al. 1995). Thus, it is evident that virus-activated T cells are equipped with the capacity to migrate to areas of inflammation; a finding which confirms and extends the classical observation that

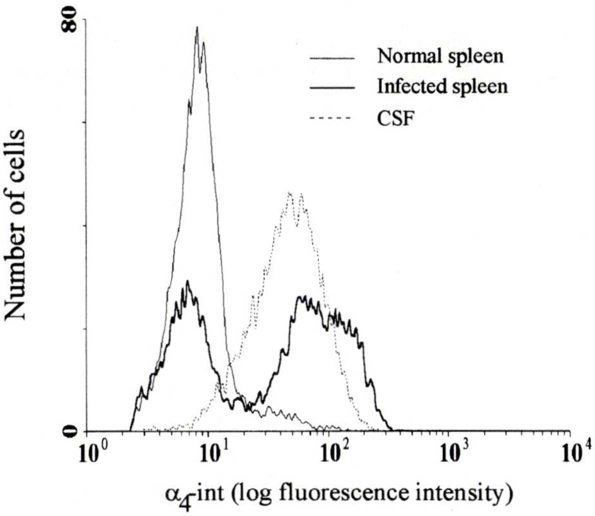

Fig. 6. Expression of α_4-integrin on $CD8^+$ cells from spleen and CSF of mice infected i.c. 7 days earlier; at this time $CD8^+$ cells constitute about half of the infiltrating cells. Gates have been set for $CD8^+$ cells and results on splenic $CD8^+$ cells from uninfected mice have been included for comparison. (Modified from ANDERSSON et al. 1994, Copyright 1994, the American Association of Immunologists)

recently activated blast cells and memory cells show a preference for homing to inflammatory sites, and that the majority of cells recruited are not antigen-specific (McCLUSKEY and WERDELIN 1971; ASHERSON and ALLWOOD 1972; NORTH and SPITALNY 1974; HURWITZ et al. 1983; ISSEKUTZ 1991).

Histological analysis of venules from infected areas – following perfusion fixation to eliminate freely circulating cells – disclosed that these contained many leukocytes sticking to the interior surfaces of the vessels. Using electron microscopy many of these leukocytes were found to closely interact with the endothelium (MARKER et al. 1984). Immunohistochemical staining for relevant ligands of the putatively involved lymphocyte receptors gave results that largely matched the expected pattern: endothelial expression of both ICAM-1 (ligand for LFA-1) and VCAM-1 (ligand for VLA-4) was found to be up-regulated as a result of the antiviral immune response (MARKER et al. 1995). However, although vessels surrounded by cells phenotypically matching activated T cells were found to be positive, venules outside of infected areas were also positive. For example, although mice infected with LCMV only develop meningeal inflammation with little infiltration of gray matter, venules deep inside the brain were more positive in infected mice than in uninfected controls. This could be taken to underscore the fact that, although expression of endothelial ligands is mandatory for lymphocyte extravasation, additional signals are needed to steer effector cell migration. Chemokines are interesting candidates in this respect (MACKAY 1996). Thus, chemokine receptors are up-regulated on activated/memory T cells (LOETSCHER et al. 1996; OIN et al. 1996), and a number of chemokines has been found in vitro to stimulate T cell migration (OIN et al. 1996; ROTH et al. 1995; CARR et al. 1994). In particular the

C-C chemokine monocyte chemoattractant protein-1 (MCP-1) has been observed to be a major T cell chemoattractant (ROTH et al. 1995), and it is therefore of interest that analysis of CSF from mice infected i.c. with LCMV has recently revealed the presence of MCP-1, the amount of which correlates with the severity of the inflamatory process (unpublished data). However, further studies are needed to clarify whether this correlation reflects that MCP-1 is a mediator of the inflammatory process or is a result thereof.

For a more detailed analysis of the role of adhesion molecules in directing T cells and other effector cells to sites of viral infection, virus-induced delayed-type hypersensitivity (DTH), elicited by local infection of the footpad with live virus,

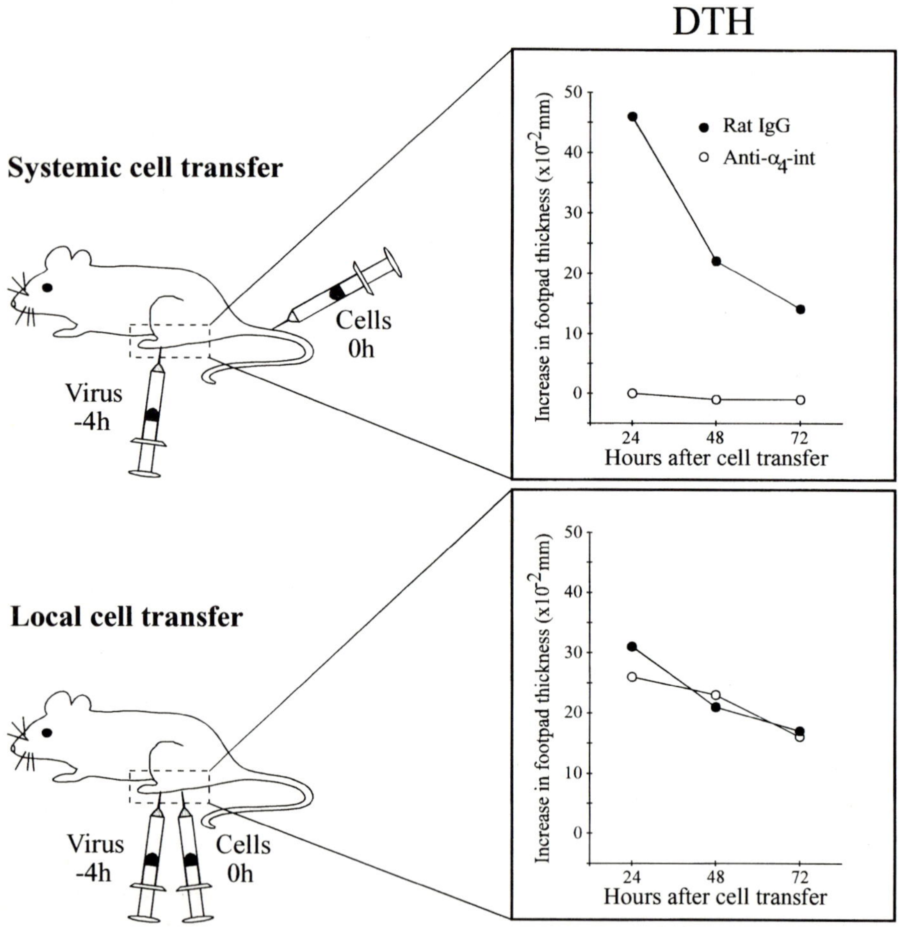

Fig. 7. Model system to evaluate the role of adhesion molecules in targeting virus-primed T cells to sites of viral replication: a typical result obtained using anti-α_4-integrin antibody. Virus-primed mitomycin C-treated donor splenocytes were preincubated with the relevant antibody and injected either i.v. or directly into the foot-pad of recipients infected in the foot 4 h prior to cell transfer. (modified from CHRISTENSEN et al. 1995, Copyright 1995, the American Association of Immunologists)

was used as model system (Fig. 7). Like the inflammatory reaction in infected organs this is a complex process involving several cell types (Rosen et al. 1989; Volkman and Collins 1971; Lubaroff and Waksman 1968). However, by studying DTH elicted by adoptive transfer of adherent-cell depleted, virus-primed effector cells, the reaction could be broken down into two major components: virus-primed, donor-derived $CD8^+$ effector cells and radiosensitive nonspecific cells delivered by the recipient (Nielsen et al. 1994). Both components were found to be required for a virus-specific DTH reaction to be elicited, since elimination of either subset reduced the reaction to background. Preincubation of the donor cells with antibodies directed against several relevant adhesion molecules revealed that anti-LFA-1 as well as anti-α_4-integrin markedly inhibited the inflammatory reaction (Andersson et al. 1995; Christensen et al. 1995). Also an antibody to CR3 significantly delayed the reaction (Nielsen et al. 1994) consistent with the finding that many of the inflammatory $CD8^+$ T cells are Mac-1^+ (McFarland et al. 1992). Neither of these antibodies inhibited the inflammatory reaction in experiments in which the requirement for effector cell homing was bypassed by directly inoculating the primed cells into the test site. This strongly indicates that the major function of the studied molecules is to direct effector cell migration to sites of infection. It should be noted that not all antibodies directed to cell surface molecules were inhibitory; thus a monoclonal antibody to the common leukocyte antigen (CD45) had little or no effect, demonstrating that mere binding to the cell surface did not suffice for inhibition of effector cell homing (Christensen et al. 1995).

Local injection of virus-primed effector cells into the test site was also used to identify the accessory cell type required for elicitation of a virus-specific inflammatory reaction. Thus preirradiation or pretreatment of the recipient with anti-CR3 completely inhibited or significantly delayed the reaction, respectively (Nielsen et al. 1994). In either case responsiveness could be restored by coinjection into the test site of peritoneal cells from unprimed mice, and depletion of adherent cells prior to transfer eliminated that potential. Thus, all evidence pointed to cells of the monocyte/macrophages lineage using CR3 as one of the major receptors for extravasation.

Whereas the results obtained by antibody blocking supported the conclusion that all three integrins demonstrated on the surface of virus-activated T cells (LFA-1, α_4-integrin, Mac-1) played a role in homing to sites of viral infection, this approach is open to criticism. That is, it could be claimed that binding of antibodies may lead to altered cellular function – especially as it is known that these molecules are capable of transducing out-to-in signals. Furthermore, these studies do not reveal the true importance of each molecule for T cell-mediated inflammation in the intact infected animal, since they were carried out using limiting numbers of effector cells in mice subjected to infection immediately prior to cell transfer. Therefore, to avoid these objections we turned to mice with targeted gene defects of relevant adhesion molecules (Sharpe 1995; Bullard et al. 1995). When these experiments were carried out, only mice deficient in expression of ICAM-1 and mice with a hypomorphic expression of CD18 (the common β_2 chain of LFA-1 and Mac-1) were available (Sligh et al. 1993; Wilson et al. 1993). Two manifestations of T cell-mediated inflammation were studied: LCMV-induced T cell-mediated meningitis and footpad swelling induced by primary infection of the footpad (Fig. 8A). Virtually no impairment of T cell-mediated inflammation was observed in ICAM-1 deficient mice and about a one day delay was noted in CD18 mutants (Christensen et al. 1996a). However, since a slight impairment of specific effector T cell generation was also found in the latter mice, it cannot be excluded that this is the main reason for the delayed inflammatory response observed. Therefore, the main conclusion of this analysis was that neither of these molecules were mandatory for the induction of T cell-mediated inflammation in virus-infected mice. In contrast, when the adoptive DTH model reaction was studied using these mutants as recipients of primed wild-type cells (Fig. 8B), the importance of ICAM-1 and CD18 in elicitation of virus-induced DTH was essentially confirmed (Christensen et al. 1996a).

Although puzzling at first sight, there may be several reasons for this apparant discrepancy between results obtained in adoptively immunized mice and intact infected animals. First, when effector cells are adoptively transferred, cells are injected as a bolus whereas in the intact mouse a continual supply of activated cells are available. Second, the state of endothelial activation may be different in mice infected immediately prior to cell transfer compared to mice 4–6 days into the infection (Irani and Griffin 1996). Therefore, functional redundancy is likely to be more pronounced in intact mice, whereas in the adoptive situation any lowering of the avidity of the T cell/endothelial cell interaction is important. Thus, all the molecules revealed to be of importance based on adoptive transfer probably are involved also in the intact animal, but redundancy in function of LFA-1/ICAM-1 and VLA-4/VCAM-1 precludes either set from being critical in the latter case. That these receptor/ligand pairs may exert overlapping functions in T cell homing is supported not only in the literature (Issekutz 1992), but also by our own studies. Thus, ICAM-1 deficient mice given wild-type effector cells have a reduced DTH reaction which, however, can be completely abolished by preincubation of the donor cells with a soluble VCAM-1 construct; if similarly treated donor cells are given to wild-type recipients, only a partial inhibition is observed (unpublished data).

8 Role of Cytokines

A critical level of regulation of T cell extravasation, aside from expression of appropriate adhesion molecules on the surface of circulating T cells, is the expression of corresponding ligands on the endothelium. Local up-regulation of the expression of these molecules serves to mark areas of interest to the generated effector cells, and thus specifically targets effector cell homing to infectious foci. This may be particularly important early in the infection when numbers of effector cells are limited. Consistent with this, up-regulation of ICAM-1 and VCAM-1 expression is found at sites of virus-induced inflammation. The signals inducing the

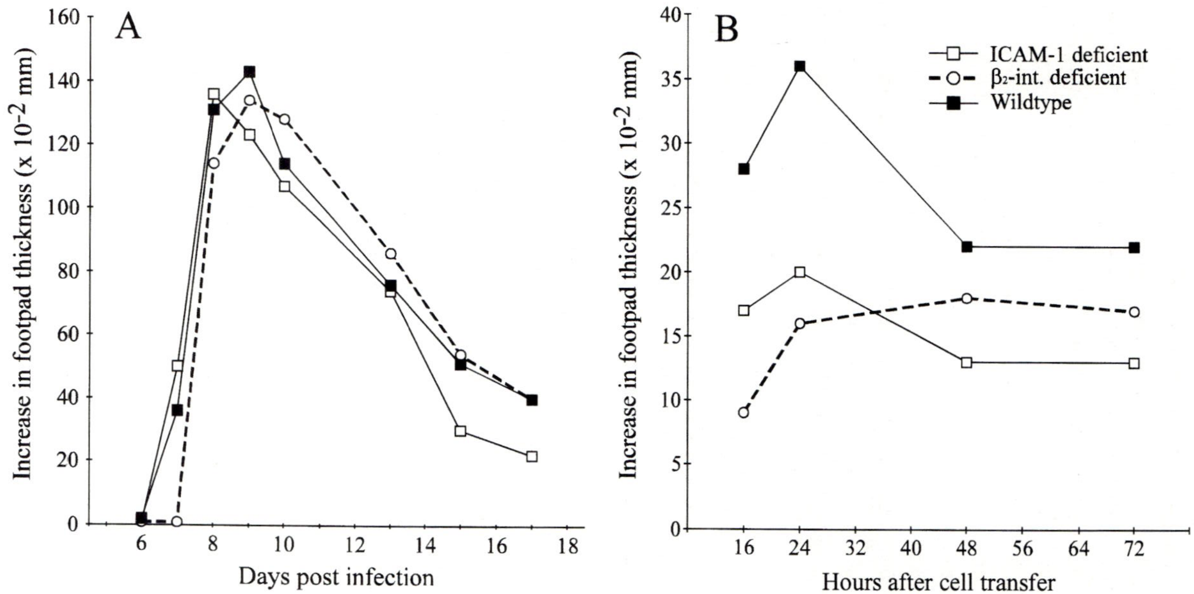

Fig. 8A, B. Effect of deficient ICAM-1 or β_2-integrin expression on primary footpad swelling (**A**) and capacity to support an adoptive footpad swelling reaction (**B**); donor cells for adoptive transfer were lymphocytic choriomeningitis virus (LCMV)-primed, wild-type cells. (From Christensen et al. 1996a)

expression of these molecules may often be cytokines induced by the virus infection per se (Irani and Griffin 1996). However, at least in the case of i.c. infection with LCMV, the inflammatory signal generated by the virus itself is very modest (Campbell et al. 1994), and in virus-infected T cell deficient mice we could not detect any up-regulation of either ICAM-1 or VCAM-1 (Marker et al. 1995). Based on the fact that IFN-γ is present in the CSF of virus-infected, immunocompetent mice but not of infected, T cell deficient mice (Frei et al. 1988), and that IFN-γ is known to up-regulate these molecules, we have proposed that T cells may regulate endothelial expression of these ligands through production of IFN-γ (Marker et al. 1995). A number of additional findings consistent with this model have recently been obtained. First, T cells found at the inflammatory site produce high amounts of IFN-γ, reflecting the focussing of activated effector cells to the inflammatory site (Christensen et al. 1996c). Second, $CD8^+$ T cells are the major inducers of LCMV-specific DTH and suffices for up-regulation of ICAM-1 and VCAM-1 expression (Nielsen et al. 1994; Christensen et al. 1994; Leist et al. 1987; Marker et al. 1995), and $CD8^+$ T cells are also the major producers of IFN-γ (Christensen et al. 1996c). In contrast, another candidate proinflammatory cytokine, TNF-α, is neither present in the CSF nor produced by LCMV-activated T cells (Leist et al. 1988; Christensen et al. 1996c). Therefore, to confirm a central role of IFN-γ, IFN-γ deficient mice were infected with LCMV and T cell-mediated meningitis was studied. The LCMV strain used was a neurotropic variant; this was chosen to minimize any effect of redistribution of virus replication on effector cell homing (Leist et al. 1989). Much to our surprise little or no difference either in T cell-mediated inflammation or in the up-regulation of ICAM-1 or VCAM-1 on meningeal venules was observed between IFN-γ deficient mice and matched infected wild-type animals (manuscript in preparation). Thus, at present the simplest interpretation appears to be that even though IFN-γ are likely to play a role under normal conditions (Campbell et al. 1994), other T cell signals may suffice for significant up-regulation of these adhesion molecules. Such redundancy may be a parallel to the redundancy observed for adhesion molecules.

9 An Integrated Model for Effector T Cell Homing to Sites of Viral Replication

Extrapolating from the results that we have obtained primarily in LCMV-infected mice, we propose the following model for the events that occur when the host responds with a T cell dependent inflammatory reaction to a viral infection. In the uninfected host, an intact vascular barrier seem to preclude effector T cell access to many organs (Ando et al. 1994a). Therefore marked changes in the interaction of activated T cell and endothelium localized in areas of infection is critical for the formation of the inflammatory exudate and thus the capacity of effector T cells to reach virus-infected cells outside of lymphoid organs.

As a result of the immune response, effector T cells are generated which are endowed with the adhesion molecules that target these cells to infected areas. It is of interest to note that a number of molecules with this potential are up-regulated on primed T cells, thus besides VLA-4 (α_4-integrin) and LFA-1, also Pgp-1 and, on many $CD8^+$ T cells, Mac-1. All of these molecules seem to be of importance for lymphocyte extravasation, and this apparant redundancy suggests that the migration of effector cells to inflamed areas is of such vital importance that even if one receptor/ligand pair is blocked, the formation of the inflammatory exudate should not be impaired. From an evolutionary perspective the extensive redundancy thus serves to reduce the risk that viruses or microorganisms acquire the capacity to substantially inhibit the inflammatory response.

Locally, expression of the adhesion molecules ICAM-1 and VCAM-1 is up-regulated, a process that is controlled by cytokines. If the virus itself causes sufficient induction of proinflammatory cytokines, this will directly allow the focussing of effector cells onto the area (Irani and Griffin 1996). In case no such signal is induced, the first effector T cells will leave the circulation at random, probably through interaction with constitutively expressed ligands, e.g., ICAM-2. Due to their higher expression of the corresponding receptors recently activated cells will dominate at this stage. Once a virus-specific T cell interacts with virus-infected cells in an infectious focus, the cytokine cascade will be triggered involving both T cell-produced cytokines (e.g., IFN-γ) and monokines released from activated monocytes/macrophages (e.g., TNF-α, MCP-1) (Nielsen et al. 1994). Together with chemokines these mediators will trigger invasion by the bulk of the effector cells, and at this stage perhaps even some nonactivated cells may be recruited through the highly activated endothelium (Doherty et al. 1988). Once the infection is controlled, most of the recruited cells probably die in situ, and memory cells generated in the lymphoid organs accumulate predominantly in the spleen due to their low expression of L-selectin early after antigenic stimulation (the precise relationship between primary effector cells and memory cells is still a controversial subject for which the reader is referred to a recent review by Ahmed and Gray 1996). However, reintroduction of virus leads to rapid mobilization of these cells from the spleen and their migration to sites of infection as well as to the lymph nodes draining such areas (unpublished data).

10 Implications

The fact that viral infections induce a high number of $CD8^+$ T cells to become activated, thus leading to changes in both effector capacity and migration pattern, clearly raises the issue of the functional consequences of this extensive T cell activation. First, is bystander activation simply an unavoidable byproduct of a triggering cascade developed to secure rapid generation of virus-specific effector T cells or are the activated nonspecific T cells of direct importance in virus control? From our results it is clear that the latter cells possess a surface phenotype that targets them to infectious foci. Furthermore, many possess qualities of relevance to combating virally infected cells, i.e., cytotoxicity and capacity to produce IFN-γ. It is tempting to speculate, therefore, that these cells could somehow contribute to virus control orchestrated by those few effector T cells that are virus-specific. For example, cognate interactions established through the adhesion molecules might lead to cell killing and/or release of IFN-γ which would reduce viral spreading, activate macrophages and increase surface expression of relevant ligands (e.g., ICAM-1 and MHC molecules). Evidently such potent T cell activation also poses a great risk to the host; nonspecific cell damage and tissue destruction being the most obvious (Ando et al. 1993, 1994b). However, the extensive generation of T cells with the above functional qualities might also initiate graft rejection (Pouteil-Noble et al. 1993) and result in autoreactivity directed towards otherwise ignored antigens; in predisposed hosts overt autoimmune disease may be the final outcome (Lee et al. 1995; Rabinovitch et al. 1995; Liblau et al. 1995; von Herrath and Oldstone 1997). Thus, epidemiologic data indicate that clinical exacerbations of multiple sclerosis are often preceded by viral infections (Sibley et al. 1985), and our findings may provide a mechanism to understand this correlation without necessarily invoking molecular mimicry between viral epitopes and the involved autoantigen(s) (illustrated in Fig. 9). Although LCMV infection has not been found to induce autoimmunity in normal mice, it has recently been observed that another viral infection (Semliki forest virus) may facilitate murine experimental allergic en-

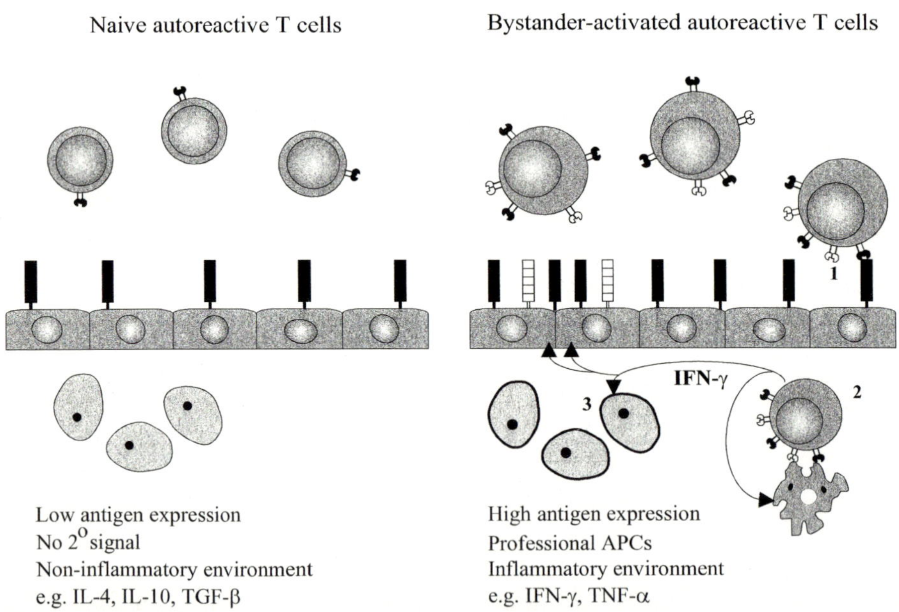

Fig. 9. A hypothetical scheme outlining the possible mechanisms whereby virus-induced bystander activation could play a role in elicitation of autoimmune reactions

cephalitis through what appears to be bystander activation and enhanced leukocyte entry into the CNS (Soilu-Hänninen et al. 1996). Notably, this facilitation could be blocked by anti-α_4-integrin monoclonal antibody therapy. Therefore, further studies are clearly needed to clarify the relevance of virus-induced bystander activation in the pathogenesis of autoimmune diseases.

Acknowledgements. The authors thank G. Thørner Andersen for expert technical assistance, and O. Marker for constructive criticism during preparation of the manuscript. J. Pravsgaard Christensen is the recipient of a Ph.D. scholarship from the Faculty of Health Sciences, University of Copenhagen.

References

Ahmed R, Gray D (1996) Immunological memory and protective immunity: understanding their relation. Science 272:54–60

Allan JE, Dixon JE, Doherty PC (1987) Nature of the inflammatory process in the central nervous system of mice infected with lymphocytic choriomeningitis virus. In: Oldstone MB (ed) Arenaviruses. Biology and immunotherapy. Springer, Berlin Heidelberg New York, pp 131–143 (Current topics in microbiology and immunology, vol 134)

Andersen IH, Marker O, Thomsen AR (1990) Breakdown of blood-brain barrier in the murine lymphocytic choriomeningitis virus infection mediated by virus-specific $CD8^+$ T cells. J Neuroimmunol 31:155–163

Andersson EC, Christensen JP, Marker O, Thomsen AR (1994) Changes in cell adhesion molecule expression on T cells associated with systemic virus infection. J Immunol 152:1237–1245

Andersson EC, Christensen JP, Scheynius A, Marker O, Thomsen AR (1995) Lymphocytic choriomeningitis virus infection is associated with long-standing perturbation of LFA-1 expression on $CD8^+$ T cells. Scand J Immunol 42:110–118

Ando K, Moriyama T, Guidotti LG, Wirth S, Schreiber RD, Schlicht HJH et al (1993) Mechanisms of class I restricted immunopathology. A transgenic mouse model of fulminant hepatitis. J Exp Med 178:1541–1554

Ando K, Guidotti LG, Cerny A, Ishikawa T, Chisari FV (1994a) CTL access to tissue antigen is restricted in vivo. J Immunol 153:482–488

Ando K, Guidotti LG, Wirth S, Ishikawa T, Missale G, Moriyama T et al (1994b) Class I–restricted cytotoxic T lymphocytes are directly cytopathic for their target cells in vivo. J Immunol 152: 3245–3253

Arbones ML, Ord DC, Ley K, Ratech H, Maynard-Curry C, Otten G et al (1994) Lymphocyte homing and leukocyte rolling and migration are impaired in L-selectin-deficient mice. Immunity 1:247–260

Asherson GL, Allwood GG (1972) Inflammatory lymphoid cells: cells in immunized lymph nodes that move to sites of infection. Immunology 22:493–502

Bachmann MF, Sebzda E, Kundig TM, Shahinian A, Speiser DE, Mak TW et al (1996) T cell responses are governed by avidity and co-stimulatory thresholds. Eur J Immunol 26:2017–2022

Beverley PC (1990) Is T cell memory maintained by cross reactive stimulation? Immunol Today 11: 203–205

Bradley LM, Duncan DD, Tonkonogy S, Swain SL (1991) Characterization of antigen-specific $CD4^+$ effector T cells in vivo: immunization results in a transient population of MEL-14-, CD45RB- helper cells that secretes interleukin 2 (IL-2), IL-3, IL-4 and interferon-gamma. J Exp Med 174:547–559

Bradley LM, Watson SR, Swain SL (1994) Entry of naive CD4 T cells into peripheral lymph nodes requires L-selectin. J Exp Med 180:2401–2406

Brod SA, Purvee M, Benjamin D, Hafler DA (1990) T-T cell interactions are mediated by adhesion molecules. Eur J Immunol 20:2259–2268

Bullard DC, Sandberg ET, Scharffetter-Kochanek K, Beaudet AL (1995) Gene targeting for inflammatory cell adhesion molecules. Agents Actions [Suppl] 47:143–154

Butcher EC (1991) Leukocyte-edothelial cell recognition: three (or more) steps to specificity and diversity. Cell 67:1033–1036

Butcher EC, Picker LJ (1996) Lymphocyte homing and homeostasis. Science 272:60–66

Camp RL, Scheynius A, Johansson C, Pure E (1993) CD44 is necessary for optimal contact allergic responses but is not required for normal leukocyte extravasation. J Exp Med 178:497–507

Campbell IL, Hobbs MV, Kemper P, Oldstone MB (1994) Cerebral expression of multiple cytokine genes in mice with lymphocytic choriomeningitis. J Immunol 152:716–723

Carp RI, Davidson AL, Merz PA (1971) A method for obtaining cerebrospinal fluid from mice. Res Vet Sci 12:499

Carr MW, Roth SJ, Luther E, Rose SS, Springer TA (1994) Monocyte chemoattractant protein 1 acts as a T-lymphocyte chemoattractant. Proc Natl Acad Sci USA 91:3652–3656

Ceredig R, Allan JE, Tabi Z, Lynch F, Doherty PC (1987) Phenotypic analysis of the inflammatory exudate in murine lymphocytic choriomeningitis. J Exp Med 165:1539–1551

Christensen JP, Marker O, Thomsen AR (1994) The role of $CD4^+$ T cells in cell-mediated immunity to LCMV. Studies in MHC class I and MHC class II deficient mice. Scand J Immunol 40:373–382

Christensen JP, Andersson EC, Scheynius A, Marker O, Thomsen AR (1995) Alpha-4 integrin directs virus-activated $CD8^+$ T cells to sites of infection. J Immunol 154:5293–5301

Christensen JP, Marker O, Thomsen AR (1996a) T cell-mediated immunity to lymphocytic choriomeningitis virus in β_2-integrin (CD18)- and ICAM-1 (CD54)-deficient mice. J Virol 70:8997–9002

Christensen JP, Röpke C, Thomsen AR (1996b) Virus-induced polyclonal T cell activation is followed by apoptosis: partitioning of $CD8^+$ T cells based on α4 integrin expression. Int Immunol 8:707–715

Christensen JP, Stenvang JP, Marker O, Thomsen AR (1996c) Characterization of virus-primed $CD8^+$ T cells with a type 1 cytokine profile. Int Immunol 8:1453–1461

Colle JH, Saron MF, Shidani B, Lembezat MP, Truffa-Bachi P (1993) High frequency of T lymphocytes committed to interferon-gamma transcription upon polyclonal activation in spleen from lymphocytic choriomeningitis virus-infected mice. Int Immunol 5:435–441

Cousens LP, Orange JS, Biron CA (1995) Endogenous IL-2 contributes to T cell expansion and IFN-γ production during lymphocytic choriomeningitis virus infection. J Immunol 155:5690–5699

Doherty PC (1973) Quantitative studies of the inflammatory process in fatal viral meningoencephalitis. Am J Pathol 73:607–622

Doherty PC, Zinkernagel RM (1974) T cell-mediated immunopathology in viral infections. Transplant Rev 19:89–120

Doherty PC, Allan JE, Ceredig R (1988) Contribution of host and donor T cells to the inflammatory process in murine lymphocytic choriomengitis. Cell Immunol 116:475–481

Doherty PC, Allan JE, Lynch F, Ceredig R (1990) Dissection of an inflammatory process induced by $CD8^+$ T cells. Immunol Today 11:55–59

Doherty PC, Allan W, Eichelberger M, Carding SR (1992) Roles of alpha beta and gamma delta T cell subsets in viral immunity. Annu Rev Immunol 10:123–151

Doherty PC, Hou S, Southern PJ (1993) Lymphocytic choriomeningitis virus induces a chronic wasting disease in mice lacking class I major histocompatibility complex glycoproteins. J Neuroimmunol 46:11–17

Doherty PC, Hou S, Tripp RA (1994) $CD8^+$ T cell memory to viruses. Curr Open Immunol 6: 545–552

Frei K, Leist TP, Meager A, Gallo P, Leppert D, Zinkernagel RM, et al (1988) Production of B cell stimulatory factor–2 and interferon gamma in the central nervous system during viral meningitis and encephalitis. Evaluation in a murine model infection and in patients. J Exp Med 168:449–453

Hou S, Doherty PC (1993) Partitioning of responder $CD8^+$ T cells in lymph node and lung of mice with Sendai virus pneumonia by LECAM-1 and CD45RB phenotype. J Immunol 150:5494–5500

Hurwitz JL, Korngold R, Doherty PC (1983) Specific and nonspecific T cell recruitment in viral meningitis: possible implications for autoimmunity. Cell Immunol 76:397–401

Irani DN, Griffin DE (1996) Regulation of lymphocyte homing into the brain during viral encephalitis at various stages of infection. J Immunol 156:3850–3857

Issekutz TB (1990) Effects of six different cytokines on lymphocyte adherence to microvascular endothelium and in vivo lymphocyte migration in the rat. J Immunol 144:2140–2146

Issekutz TB (1991) Effect of antigen challenge on lymph node lymphocyte adhesion to vascular endothelial cells and the role of VLA-4 in the rat. Cell Immunol 138:300–312

Issekutz TB (1992) Lymphocyte homing to sites of inflammation. Curr Open Immunol 4:287–293

Jung TM, Gallatin WM, Weissman IL, Dailey MO (1988) Down-regulation of homing receptors after T cell activation. J Immunol 141:4110–4117

Kagi D, Ledermann B, Burki K, Seiler P, Odermatt B, Olsen KJ et al (1994) Cytotoxicity mediated by T cells and natural killer cells is greatly impaired in perforin-deficient mice. Nature 369:31–37

Kasaian MT, Biron CA (1989) The activation of IL-2 transcription in L3T4$^+$ and Lyt-2$^+$ lymphocytes during virus infection in vivo. J Immunol 142:1287–1292

Kasaian MT, Leite-Morris KA, Biron CA (1991) The role of CD4$^+$ cells in sustaining lymphocyte proliferation during lymphocytic choriomeningitis virus infection. J Immunol 146:1955–1963

Kramer MD, Fruth U, Simon H, Simon MM (1989) Expression of cytoplasmic granules with T cell associated serine proteinase-1 activity in Ly-2+ (CD8$^+$) T lymphocytes responding to lymphocytic choriomeningitis virus in vivo. Eur J Immunol 19:151–156

Kündig TM, Althage A, Hengartner H, Zinkernagel RM (1992) Skin test to assess virus-specific cytotoxic T cell activity. Proc Natl Acad Sci USA 89:7757–7761

Kündig TM, Bachmann MF, DiPaolo C, Simard JJ, Battegay M, Lother H et al (1995) Fibroblasts as efficient antigen-presenting cells in lymphoid organs. Science 268:1343–1347

Lau LL, Jamieson BD, Somasundaram T, Ahmed R (1994) Cytotoxic T cell memory without antigen. Nature 369:648–652

Lee MS, von Herrath M, Reiser H, Oldstone MB, Sarvetnick N (1995) Sensitization to self (virus) antigen by in situ expression of murine interferon-gamma. J Clin Invest 95:486–492

Lehmann-Grube F, Lohler J, Utermohlen O, Gegin C (1993) Antiviral immune responses of lymphocytic choriomeningitis virus infected mice lacking CD8$^+$ T lymphocytes because of disruption of the beta 2-microglobulin gene. J Virol 67:332–339

Leist T, Cobbold SP, Waldmann H, Aguet M, Zinkernagel RM (1987) Functional analysis of T lymphocyte subset in antiviral host defence. J Immunol 138:2278–2281

Leist TP, Frei K, Kam-Hansen S, Zinkernagel RM, Fontana A (1988) Tumor necrosis factor alpha in cerebrospinal fluid during bacterial, but not viral, meningitis. Evaluation in murine model infections and in patients. J Exp Med 167:1743–1748

Leist TP, Eppler M, Zinkernagel RM (1989) Enhanced virus replication and inhibition of lymphocytic choriomeningitis virus disease in anti-gamma interferon-treated mice. J Virol 63:2813–2819

Liblau RS, Singer SM, McDevitt HO (1995) Th1 and Th2 CD4$^+$ T cells in the pathogenesis of organ-specific autoimmune diseases. Immunol Today 16:34–38

Loetscher P, Seitz M, Baggiolini M, Moser B (1996) Interleukin-2 regulates CC chemokine receptor expression and chemotactic responsiveness in T lymphocytes. J Exp Med 184:569–577

Lubaroff DM, Waksman BH (1968) Bone marrow as a source of cells in reactions of cellular hypersensitivity. II. Identification of allogeneic cells or hybrid cells by immunofluorescence in passively transferred tuberculin reactions. J Exp Med 128:1437–1447

Lynch F, Doherty PC, Ceredig R (1989) Phenotypic and functional analysis of the cellular response in regional lymphoid tissue during an acute virus infection. J Immunol 142:3592–3598

Mackay CR (1991) T cell memory: the connection between function, phenotype and migration pathways. Immunol Today 12:189–192

Mackay CR (1996) Chemokine receptors and T cell chemotaxis. J Exp Med 184:799–802

Mackay CR, Marston W, Dudler L (1992) Altered patterns of T cell migration through lymph nodes and skin following antigen challenge. Eur J Immunol 22:2205–2210

Marker O, Volkert M (1973) Studies on cell-mediated immunity to lymphocytic choriomeningitis virus in mice. J Exp Med 137:1511–1525

Marker O, Nielsen MH, Diemer NH (1984) The permeability of the blood-brain barrier in mice suffering from fatal lymphocytic choriomeningitis virus infection. Acta Neuropathol (Berl) 63:229–239

Marker O, Scheynius A, Christensen JP, Thomsen AR (1995) Virus-activated T cells regulate expression of adhesion molecules on endothelial cells in sites of infection. J Neuroimmunol 62:35–42

Matzinger P (1994) Tolerance, danger, and the extended family. Annu Rev Immunol 12:991–1045

McCluskey RT, Werdelin O (1971) The origin and nature of the mononuclear cells participating in immunologically induced inflammatory reactions. In: Cudkowicz CS, McCluskey G (eds) Cellular interactions in the immune response. Karger, Basel

McFarland HI, Nahill SR, Maciaszek JW, Welsh RM (1992) CD11b (Mac-1): a marker for CD8$^+$ cytotoxic T cell activation and memory in virus infection. J Immunol 149:1326–1333

McIntyre KW, Welsh RM (1986) Accumulation of natural killer and cytotoxic T large granular lymphocytes in the liver during virus infection. J Exp Med 164:1667–1681

Mobley JL, Dailey MO (1992) Regulation of adhesion molecule expression by CD8$^+$ T cells in vivo: differential regulation of gp90$^{Mel\ 14}$ (LECAM-1), Pgp-1, LFA-1, and VLA-4α during the differentiation of cytotoxic T lymphocytes induced by allografts. J Immunol 148:2348–2356

Mobley JL, Rigby SM, Dailey MO (1994) Regulation of adhesion molecule expression by CD8 T cells in vivo. II. Expression of L-selectin (CD62L) by memory cytolytic T cells responding to minor histocompatibility antigens. J Immunol 153:5443–5452

Moskophidis D, Cobbold SP, Waldmann H, Lehmann-Grube F (1987) Mechanism of recovery from acute virus infection: treatment of lymphocytic choriomeningitis virus-infected mice with monoclonal antibodies reveals that Lyt-2+ T lymphocytes mediate clearance of virus and regulate the antibody response. J Virol 61:1867–1874

Moskophidis D, Fang L, Grossmann J, Drjupin R, Löhler J, Bruns M et al (1990) Virus-specific delayed-type hypersensitivity (DTH): cells mediating lymphocytic choriomeningitis virus-specific DTH reaction in mice. J Immunol 144:1926–1934

Nahill SR, Welsh RM (1993) High frequency of cross-reactive cytotoxic T lymphocytes elicited during the virus-induced polyclonal cytotoxic T lymphocyte response. J Exp Med 177:317–327

Nielsen HV, Christensen JP, Andersson EC, Marker O, Thomsen AR (1994) Expression of type 3 complement receptor on activated $CD8^+$ T cells facilitates homing to inflammatory sites. J Immunol 153:2021–2028

North RJ, Spitalny G (1974) Inflammatory lymphocytes in cell-mediated antibacterial immunity: factors governing the accumulation of mediator T cells in peritoneal exudates. Infect Immun 10:489–498

Oin S, LaRosa G, Campbell JJ, Smith-Heath H, Kassam N, Shi X et al (1996) Expression of monocyte chemoattractant protein-1 and interleukin-8 receptors on subsets of T cells: correlation with trans-endothelial chemotactic potential. Eur J Immunol 26:640–647

Pfau CJ, Thomsen AR (1993) Lymphocytic choriomeningitis virus. History of the gold standard for viral immunobiology. In: Salvato MS (ed) The arenaviridae. Plenum, New York, pp 191–198

Pouteil-Noble C, Ecochard R, Landrivon G, Donia-Maged A, Tardy JC, Bosshard S et al (1993) Cytomegalovirus infection – an etiological factor for rejection? A prospective study in 242 renal transplant patients. Transplantation 55:851–857

Rabinovitch A, Suarez-Pinzon WL, Sorensen O, Bleackley RC, Power RF (1995) IFN-gamma gene expression in pancreatic islet-infiltrating mononuclear cells correlates with autoimmune diabetes in nonobese diabetic mice. J Immunol 154:4874–4882

Razvi ES, Jiang Z, Woda BA, Welsh RM (1995a) Lymphocyte apoptosis during the silencing of the immune response to acute viral infections in normal, lpr, and Bcl-2-transgenic mice. Am J Pathol 147:79–91

Razvi ES, Welsh RM, McFarland HI (1995b) In vivo state of antiviral CTL precursors. Characterization of a cycling cell population containing CTL precursors in immune mice. J Immunol 154:620–632

Rosen H, Milon G, Gordon S (1989) Antibody to the murine type 3 complement receptor inhibits T lymphocute-dependent recruitment of myelomonocytic cells in vivo. J Exp Med 169:535–548

Roth MD (1994) Interleukin 2 induces the expression of CD45RO and the memory phenotype by CD45RA+ peripheral blood lymphocytes. J Exp Med 179:857–864

Roth SJ, Carr MW, Springer TA (1995) C-C chemokines, but not the C-X-C chemokines interleukin-8 and interferon-gamma inducible protein-10, stimulate transendothelial chemotaxis of T lymphocytes. Eur J Immunol 25:3482–3488

Selin LK, Nahill SR, Welsh RM (1994) Cross-reactivities in memory cytotoxic T lymphocyte recognition of heterologous viruses. J Exp Med 179:1933–1943

Sharpe AH (1995) Analysis of lymphocyte costimulation in vivo using transgenic and "knockout" mice. Curr Open Immunol 7:389–395

Shimizu Y, Newman W, Tanaka Y, Shaw S (1992) Lymphocyte interaction with endothelial cells. Immunol Today 13:106–112

Sibley WA, Bamford CR, Clark K (1985) Clinical viral infections and multiple sclerosis. Lancet 1: 1313–1315

Sligh JE, Ballantyne CM, Rich SS, Hawkins HK, Smith CW, Bradley A et al (1993) Inflammatory and immune responses are impaired in mice deficient in intercellular adhesion molecule 1. Proc Natl Acad Sci USA 90:8529–8533

Soilu-Hänninen M, Röyttä M, Salmi A, Salonen R (1996) Therapy with antibody against leukocyte integrin VLA-4 is effective and safe in virus facilitated EAE (abstract). Scand J Immunol 43:727

Springer TA (1990) Adhesion receptors of the immune system. Nature 346:425–434

Springer TA (1994) Traffic signals for lymphocyte recirculation and leukocyte emigration: the multistep paradigm. Cell 76:301–314

Tabi Z, Lynch F, Ceredig R, Allan JE, Doherty PC (1987) Virus-specific memory T cells are Pgp-1+ and can be selectively activated with phorbol ester and calcium ionophore. Cell Immunol 113:268–277

Thomsen AR, Marker O (1989) MHC and non-MHC genes regulate elimination of lymphocytic choriomeningitis virus and antiviral cytotoxic T lymphocyte and delayed-type hypersensitivity mediating T lymphocyte activity in parallel. J Immunol 142:1333–1343

Tough DF, Borrow P, Sprent J (1996) Induction of bystander T cell proliferation by viruses and type I interferon in vivo. Science 272:1947–1950

Tripp RA, Hou S, Doherty PC (1995) Temporal loss of the activated L-selectin-low phenotype for virus-specific $CD8^+$ memory T cells. J Immunol 154:5870–5875

Volkert M, Lundstedt C (1968) The provocation of latent lymphocytic choriomeningitis virus infections in mice by treatment with antilymphocytic serum. J Exp Med 127:327–339

Volkman A, Collins FM (1971) The resorative effect of peritoneal macrophages on delayed hypersensitivity following ionizing radiation. Cell Immunol 2:552–566

von Herrath MG, Oldstone MBA (1997) Interferon-gamma is essential for destruction of β cells and development of insulin-dependent diabetes mellitus. J Exp Med 185:531–539

Vyth-Dreese FA, Dellemijn TAM, Frijhoff A, van Kooyk Y, Frigor CG (1993) Role of LFA-1/ICAM-1 in interleukin-2-stimulated lymphocyte proliferation. Eur J Immunol 23:3292–3299

Wilson RW, Ballantyne CM, Smith CW, Montgomery C, Bradley A, O'Brien WE et al (1993) Gene targeting yields a CD18-mutant mouse for study of inflammation. J Immunol 151:1571–1578

Xu J, Grewal IS, Geba GP, Flavell RA (1996) Impaired primary T cell responses in L-selectin-deficient mice. J Exp Med 183:589–598

Yang H, Dundon PL, Nahill SR, Welsh RM (1989) Virus-induced polyclonal cytotoxic T lymphocyte stimulation. J Immunol 142:1710–1718

Zimmerman C, Brduscha-Riem K, Blaser C, Zinkernagel RM, Pircher H (1996) Visualization, characterization, and turnover of $CD8^+$ memory T cells in virus-infected hosts. J Exp Med 183:1367–1375

Zinkernagel RM, Doherty PC (1973) Cytotoxic thymus-derived lymphocytes in cerebrospinal fluid of mice with lymphocytic choriomeningitis. J Exp Med 138:1266–1269

Zinkernagel RM, Haenseler E, Leist T, Cerny A, Hengartner H, Althage A (1986) T cell-mediated hepatitis in mice infected with lymphocytic choriomeningitis virus. Liver cell destruction by H-2 class I-restricted virus-specific cytotoxic T cells as a physiological correlate of the 51Cr-release assay? J Exp Med 164:1075–1092

α4 Integrins and Tumor Metastasis

B. Holzmann, U. Gosslar, and M. Bittner

1 Introduction . . . 125
2 Distribution of α4-Integrins . . . 126
2.1 Distribution on Normal Cells . . . 126
2.2 Distribution on Human Tumor Cells . . . 126
3 Ligand Binding Functions of α4-Integrins . . . 128
4 Regulation of Integrin Ligand Binding Activity . . . 128
5 Role of α4-Integrins for Leukocyte Activation and Apoptosis . . . 129
6 Diverse Functions of α4-Integrins for Distinct Steps of Metastasis Formation . . . 130
6.1 Regulation of Tumor Cell Adhesion to Endothelium by α4-Integrins . . . 130
6.2 Regulation of Tumor Cell Invasiveness by α4-Integrins . . . 133
6.3 Expansion of Disseminated Lymphoid Tumor Cells Is Inhibited by α4-Integrins . . . 135
7 Summary . . . 137
References . . . 138

1 Introduction

Neoplastic cells selected to form metastatic tumors pass through a cascade of events that are initiated by detachment from the primary tumor mass and invasion of the adjacent tissue (Liotta et al. 1991; Nicolson 1982; Stetler-Stevenson et al. 1993). The migration through the interstitial stroma, mainly comprised of extracellular matrix proteins such as vitronectin, collagens or fibronectin, is followed by entry into the vasculature, either by transport through the lymphatic vessels and lymphaticovascular connections or by active movement across the endothelial barrier. After dissemination through the circulation, which is a crucial step for the survival of metastatic cells, the tumor cells arrest in the microvasculature of distant organs as single cells, tumor cell emboli, or tumor cell-platelet emboli via interaction with vascular endothelium or subendothelial basement membrane. After extravasation and invasion of the target organ disseminated cells expand and generate secondary tumors.

Institute of Medical Microbiology, Immunology, and Hygiene, Technical University, Trogerstraße 9, 81675 Munich, Germany

Cell surface receptors of the integrin family mediate adhesion of tumor cells to one another, to heterologous cells, or to matrix proteins. Integrins may therefore control the development of metastatic tumors by regulating the adhesive capacity of malignant cells. In addition, integrins may influence the development of tumor metastases by ligand induced signal transduction events resulting in alterations of growth, susceptibility to apoptosis, differentiation, or proteolytic activity of tumor cells. In the present review, the role of the α4 subfamily of integrins in metastasis formation by solid tumors and lymphoid malignancies is discussed.

2 Distribution of α4-Integrins

2.1 Distribution on Normal Cells

The integrin α4 subunit is noncovalently associated with either β1 or β7 chains. Both α4β1 and α4β7 integrins are expressed on the large majority of naive $CD4^+$ and $CD8^+$ T cells (Erle et al. 1994; Schweighofer et al. 1993). Whereas the expression of α4 integrins is up-regulated on all $CD45RA^-$ memory T cells, α4β7 expression is restricted to a subset of gut associated $CD4^+$ memory T cells (Kilshaw and Murant 1990; Schweighofer et al. 1993; Shimizu et al. 1990). Peripheral blood B cells, but not resident B cells from lymphoid organs, constitutively express both α4β1 and α4β7 integrins (Postigo et al. 1993). Stimulation of resident B cells with phorbol ester for several days, however, induces α4 integrin expression (Postigo et al. 1993). In addition, both α4β1 and α4β7 integrins are expressed by natural killer cells and eosinophils (Erle et al. 1994). While α4 integrins were not detected on human neutrophils (Erle et al. 1994; Hemler et al. 1990), expression of functional levels of α4 integrins was recently described for rat neutrophils (Issekutz et al. 1996). Peripheral blood monocytes selectively express α4β1 integrin, but expression of α4β7 is up regulated upon induction of macrophage differentiation with phorbol ester or interferon (IFN)-γ (Erle et al. 1994; Tiisala et al. 1995).

2.2 Distribution on Human Tumor Cells

Expression of α4 integrins was demonstrated on many different human tumors and tumor cell lines. The results of these studies are summarized in Table 1. It was shown that α4 integrins are absent from cultured melanocytes but are detected on different cell lines derived from metastatic melanomas (Albelda et al. 1990). Consistent with these findings in situ analysis of human melanomas revealed that highly invasive primary (vertical growth phase) and metastatic melanomas expressed α4 integrins more frequently than radial growth phase primary melanomas that display a low metastatic capacity (Albelda et al. 1990). These data indicate that expression of α4 integrins on melanoma cells is up regulated during tumor

Table 1. Expression of α4-integrins on human tumor cells

Tumors		α4-Integrin expression	References
Melanoma	Primary melanoma, radial growth phase	α4β1 low	Albelda et al. 1990
	Primary melanoma, vertical growth phase	α4β1 moderate-high	Albelda et al. 1990
	Metastatic melanoma	α4β1 moderate-high	Albelda et al. 1990
Sarcoma	Primary sarcomas	α4β1 moderate α4β7 absent	Paavonen et al. 1994
	Sarcoma metastases	α4β1 high α4β7 absent	Paavonen et al. 1994
Carcinoma	Adenocarcinomas	α4β1 and α4β7 absent	Paavonen et al. 1994
Immortalized B lymphoblasts	B-lymphoblastoid cell lines (B-LCL)[a]	α4β1 high	Rincon et al. 1992
B-lineage non-Hodgkin's lymphomas	Burkitt's lymphoma (BL)[b]	α4β1 low α4β7 absent	Rincon et al. 1992; Drillenburg et al. 1997
	Nodal mantle cell lymphoma (MC)	α4β1 moderate α4β7 absent	Pals et al. 1994; Drillenburg et al. 1997
	Malignant lymphomatous polyposis (MLP)	α4β1 moderate α4β7 moderate-high	Pals et al. 1994; Drillenburg et al. 1997
T-lineage non-Hodgkin's lymphomas	Cutaneous T cell lymphoma	α4β7 absent	Drillenburg et al. 1997
	Primary mucosal T cell lymphoma	α4β7 high	Drillenburg et al. 1997

Expression of α4 integrins was determined by flow cytometry or immunohistochemistry.
[a]B lymphoblasts were immortalized in vitro by infection with Epstein-Barr virus.
[b]Type I (biopsy like, BL) cell lines as well as tumor tissues were investigated.

progression or that melanoma cells expressing elevated levels of α4 integrins are selectively expanded. Similarly, moderate levels of α4β1 integrins were found on primary sarcomas, while α4 integrins were highly expressed by metastatic sarcoma cells (Paavonen et al. 1994).

On lymphoid tumors, unique patterns of expression were documented for α4β1 and α4β7 integrins. Whereas high levels of α4β1 integrin expression were detected on B lymphoblasts immortalized in vitro by infection with Epstein-Barr virus, B lymphoma cells derived from Burkitt's tumors exhibited low α4β1 integrin expression (Rincon et al. 1992). Consistent with a function of α4β7 integrin as lymphocyte homing receptor for mucosal sites, various non-Hodgkin's lymphomas that preferentially localize to mucosa-associated lymphoid tissues show high levels of α4β7 expression (Drillenburg et al. 1997; Pals et al. 1994). By contrast, α4β7 integrins are almost absent on non-Hodgkin's lymphomas that are localized in lymph nodes or cutaneous sites (Drillenburg et al. 1997; Pals et al. 1994).

3 Ligand Binding Functions of α4-Integrins

The ligand binding specificities of α4β1 and α4β7 integrins are partially overlapping. The vascular cell adhesion molecule-1 (VCAM-1) and fibronectin are recognized by both α4β1 and α4β7 integrins with similar efficiency, while the mucosal vascular addressin MAdCAM-1 is a preferential ligand for integrin α4β7 (Berlin et al. 1993). Binding of α4β1 to MAdCAM-1 requires the high avidity state of α4β1 and is less efficient than adhesion mediated by integrin α4β7 (Strauch et al. 1994). MAdCAM-1 is a tissue-selective endothelial adhesion receptor for lymphocytes and is constitutively expressed at sites of lymphocyte extravasation into mucosal sites (Briskin et al. 1993; Streeter et al. 1988), whereas VCAM-1 is induced on endothelial cells after stimulation with various cytokines including interleukin (IL)-1, tumor necrosis factor (TNF)-α, IL-4, or IL-13 (Carlos and Harlan 1994; Masinovsky et al. 1990; Thornhill and Haskard 1990; Wellicome et al. 1990). In addition, α4β1 integrin has been reported to bind to the outer membrane protein invasin of the intracellular pathogen *Yersinia pseudotuberculosis* (Ennis et al. 1993) and to function as a receptor for thrombospondin, a matrix protein highly expressed in damaged and inflamed tissues (Mosher 1990; Yabkowitz et al. 1993). Recently, it was demonstrated that both α4β1 and α4β7 mediate interactions with isolated α4 integrin subunits in a homophilic manner (Altevogt et al. 1995).

Fibronectin contains multiple binding sites for α4β1 integrin including the CS1 and CS5 sites within the alternatively spliced type III connecting segment (Komoriya et al. 1991; Mould et al. 1990, 1991; Nojima et al. 1990), the H1 site within the Hep II domain (Mould et al. 1991), and an RGDS adhesion site that is located in the central cell binding fragment (Sanchez-Aparicio et al. 1994). Experiments using an anti-β1 monoclonal antibody that induces the activated conformation of α4β1 integrin suggest that the CS1 adhesion site is recognized by α4β1 integrin with higher affinity than the RGDS site (Sanchez-Aparicio et al. 1994). Several recent reports have shown that integrin α4β7 also recognizes the CS1 adhesion site and binds fibronectin as efficiently as α4β1 (Postigo et al. 1993; Ruegg et al. 1992; Strauch et al. 1994). In VCAM-1, binding sites for α4β1 and α4β7 have been mapped to the homologous first and fourth immunoglobulin-like domains of the seven domain form of VCAM-1 (Kilger and Holzman 1995; Kilger et al. 1997; Osborn et al. 1994; Vonderheide and Springer 1992). In MAdCAM-1, the main recognition site for integrin α4β7 is located in the first immunoglobulin-like domain (Briskin et al. 1996; Viney et al. 1996).

4 Regulation of Integrin Ligand Binding Activity

Regulation of integrin receptor activity plays a crucial role in cell adhesion to other cells or to matrix proteins. For efficient ligand binding of integrins mere expression on the cell surface is not sufficient. It is necessary that integrins are converted to an active state to bind their cognate ligands (HYNES 1992). For α4 integrins several pathways of cellular activation have been identified that lead to a rapid and transient increase of integrin activity. Importantly, these mechanisms do not affect cell surface density of α4 integrins, but appear to operate either by modulating conformation dependent integrin affinity for ligand or cytoskeleton dependent cell surface clustering of integrin receptors (HYNES 1992). It was demonstrated that cross-linking of the T cell receptor for antigen (TCR)/CD3 complex or CD2 on resting $CD4^+$ T cells induced enhanced α4β1 integrin binding to fibronectin (SHIMIZU et al. 1990). Furthermore, various chemokines including macrophage inflammatory protein (MIP)-1α, MIP-1β, RANTES and interferon inducible protein 10 (IP-10) were shown to induce binding of resting and activated T cells to recombinant VCAM-1 (LLOYD et al. 1996). Adhesion of bone marrow $CD34^+$ progenitor cells to fibronectin, present in the bone marrow stroma, depends on activation of α4β1- and α5β1-integrins in response to stimulation of stem cells with IL-3, granulocyte/macrophage colony-stimulating factor (GM-CSF) or stem cell factor (LEVESQUE et al. 1995). Interestingly, the cytokine dependent activation of hematopoietic progenitor cell adhesion was specific for α4β1 and α5β1 integrins, since other β1 integrins were not affected (LEVESQUE et al. 1995). Recently, it has been reported that expression of a constitutively active form of R-ras in a myeloid cell line enhanced binding activity to the ligand fibronectin. This effect was at least partially inhibited by peptides that specifically blocked α4β1 integrin-mediated cell adhesion (ZHANG et al. 1996).

Collectively, these results demonstrate rapid and transient regulation of α4 integrin ligand binding activity by intracellular signals that are triggered by distinct classes of cell surface receptors. Therefore, mere analysis of integrin expression levels on normal or transformed cells does not provide important information on the functional state of integrin receptors. For example, low levels of integrins locked in a high activity state could play an important role in regulating various cellular functions, while high levels of integrins that are frozen in an inactive state may represent inert cell surface components. Thus, the unique mechanisms of regulating integrin ligand binding functions pose important limitations to the interpretation of tissue distribution data.

5 Role of α4-Integrins for Leukocyte Activation and Apoptosis

T lymphocyte activation requires signals delivered by the TCR and accessory receptors. When exposed to immobilized VCAM-1 in conjunction with anti-TCR/CD3 monoclonal antibody, resting T cells can be induced to proliferate, to secrete IL-2, TNF-α, IL-4, and GM-CSF, and to increase expression of CD28 and cytotoxic T lymphocyte antigen (CTLA)-4 (BURKLY et al. 1991; DAMLE and ARUFFO 1991; DAMLE et al. 1994; UDAGAWA et al. 1996; VAN SEVENTER et al. 1991).

VCAM-1, when co-immobilized with CD3 monoclonal antibody, also induced the activation of transcription factors NF-AT, AP-1, and NF-κB (UDAGAWA et al. 1996). The role of α4 integrins in T cell costimulation have been confirmed in a rat model of immunity to the nematode *Trichinella spiralis* (BELL and ISSEKUTZ 1993). These experiments clearly demonstrate that antibodies to α4 integrins may not only inhibit protective immunity to *Trichinella spiralis* infection by blocking lymphocyte migration to the gut but also by preventing initial activation of protective $CD4^+$ T lymphocytes and by suppression of T cell effector functions after entry into the gut.

In contrast to resting or short-term activated T cells, coligation of the TCR and α4 integrins on chronically stimulated T cells results in activation-dependent death (DAMLE et al. 1993). Simultaneous stimulation of T cells with antibodies to CD2 and CD28 or exogenous addition of IL-2 and IL-4 did not reverse the death-promoting effects of α4 integrin cosignals. It therefore appears that triggering of α4 integrins may either transmit stimulatory signals or induce cell death depending on the state of activation or differentiation of T lymphocytes.

For B cell maturation and selection in germinal centers, interactions with follicular dendritic cells are of critical importance. Adhesion of B cells to follicular dendritic cells involves interactions of lymphocyte function-associated antigen (LFA)-1 with ICAM-1 and of α4 integrins with VCAM-1 (FREEDMAN et al. 1990; KOOPMAN et al. 1991). Disruption of cell clusters formed by B lymphocytes and follicular dendritic cells by monoclonal antibodies against VCAM-1, ICAM-1, α4 integrin, or LFA-1 results in apoptosis of B cells (KOOPMAN et al. 1994). Consistent with a role of α4 integrins and LFA-1 in positive selection of activated B cells in germinal centers, VCAM-1 and ICAM-1 act synergistically with anti-IgM to inhibit apoptosis of germinal center B cells (KOOPMAN et al. 1994).

6 Diverse Functions of α4-Integrins for Distinct Steps of Metastasis Formation

Numerous animal models and analyses of human tumors have established a critical role of α4 integrins for various steps during the development of metastatic tumors including detachment of cells from the primary tumor, invasion of secondary sites by circulating tumor cells and expansion or survival of tumor cells after infiltration of target organs. The results are discussed in the following chapters and are summarized in Table 2.

6.1 Regulation of Tumor Cell Adhesion to Endothelium by α4-Integrins

In animal models it was demonstrated that the metastatic capacity of melanoma cells was enhanced through α4β1-VCAM-1 interactions (GAROFALO et al. 1995; OKAHARA et al. 1994). Inflammatory cytokines that induce the expression of

Table 2. Effects of α4-integrins on experimental tumor development and metastasis

Tumor cells	Recipients		Route of injection	Local tumor growth	Metastasis formation			References
	Strain	Treatment			Lung	Spleen, lymph nodes	Bone marrow	
B16 melanoma (murine)	C57BL/6 (syngenic)	None	s.c.	+, no effect	Reduced	n.d.	n.d.	Quian et al. 1994
			i.v.	n.a.	+, no effect	n.d.	n.d.	
B16 melanoma (murine)	C57BL/6 (syngenic)	TNFα	i.v.	n.a.	Enhanced	n.d.	n.d.	Okahara et al. 1994
A375 M melanoma (human)	Nude mouse	IL-1	i.v.	n.a.	Enhanced	n.d.	n.d.	Garofalo et al. 1995
LB T-cell lymphoma (murine)	Balb/c (syngenic)	None	i.v.	n.a.	Reduced	Reduced	+, no effect	Gosslar et al. 1996
K562 erythroleukemia (human)	SCID mouse	None	s.c.	Reduced	n.d.	n.d.	n.d.	Matsuura et al. 1996
			i.v.	n.a.	n.d.	n.d.	Enhanced	
CHO fibroblasts (hamster)	SCID mouse	None	s.c.	+, no effect	–, no effect	–, no effect	–, no effect	Matsuura et al. 1996
			i.v.	n.a.	+, no effect	Enhanced	Enhanced	

Effects of α4 integrins on tumor development and metastasis formation was analyzed in different mouse models. The cell lines used were selected for overexpression of α4 integrins after cDNA transfection (B16, LB, K562, and CHO) or derived from spontaneous expression variants (A375 M). In some models recipient animals were treated with cytokines 3 h prior to tumor cell injection.
+, tumors detected; –, no tumor detectable; n.d., not determined; n.a., not applicable.

VCAM-1 on endothelial cells were shown to enhance the metastatic capacity of α4β1 positive melanoma cells (Table 2). Thus, i.v. injection of human melanoma cells into IL-1 pretreated nude mice resulted in increased numbers of lung metastases. Increased metastasis formation was inhibited by pretreatment of the tumor cells with an α4β1 integrin antibody indicating that the interaction of α4β1 integrins with VCAM-1 or the alternatively spliced CS-1 segment of fibronectin exposed on the surface of stimulated endothelium facilitated the adhesion and transmigration of melanoma cells (Garofalo et al. 1995; Elices et al. 1994). Similarly, injection of murine B16 melanoma cells into TNF-α pretreated mice resulted in a strong increase of lung metastasis formation (Table 2). Pretreatment of the tumor cells with an α4β1 integrin antibody or i.v. injection of an VCAM-1 antibody inhibited the increase in metastasis formation. These findings indicate that VCAM-1 is the preferential ligand on activated endothelium for α4β1 integrin expressed by circulating melanoma cells (Okahara et al. 1994).

In vitro binding studies revealed that different human sarcoma cell lines preferentially adhere to stimulated endothelial cells via α4β1-VCAM-1 interactions (Mattila et al. 1992; Taichman et al. 1991). The histocytochemical analysis of metastatic lesions demonstrated that VCAM-1 positive endothelial cells colocalize with sarcoma cells expressing α4β1 integrins. In addition, the metastatic lesions were located in the close vicinity of major vessels expressing VCAM-1 suggesting that sarcoma cells may have extravasated through VCAM-1 expressing endothelium (Paavonen et al. 1994). Together, these findings support the view that interaction of α4β1 integrin with the inducible vascular ligand VCAM-1 promotes sequestration of tumor cells in distant organs. In these tumor models enhanced accumulation of melanoma cells was associated with the development of an increased number of metastatic tumors suggesting that α4 integrins did not inhibit expansion of disseminated tumor cells.

In additional studies, de novo expression of α4β1 integrins was shown to enhance the metastatic capacity of Chinese hamster ovary (CHO) cells (Matsuura et al. 1996). While i.v. injection of wild-type CHO cells resulted in the formation of lung metastases exclusively, α4β1 integrin-positive CHO cells invaded additional organs including adrenals, lymph nodes and bone marrow (Table 2). The altered dissemination of CHO cells expressing integrin α4β1 might be attributed to an enhanced adhesion of CHO cells to the endothelium via α4β1 integrins. In addition, it seems possible that interactions of α4β1 integrin-positive CHO cells with fibronectin or VCAM-1 present in the stroma of target organs may retain the tumor cells that have passed the endothelial barrier.

In addition to α4β1 integrin, the localization of lymphoid tumors may be regulated by the integrin α4β7, which functions as a mucosal homing receptor (Berlin et al. 1993; Holzmann and Weissman 1989; Holzmann et al. 1989; Hu et al. 1992). Analysis of tumors derived from patients with malignant lymphomatous polyposis (MLP) indicated that the interaction of α4β7 integrins with the vascular counter receptor MAdCAM-1 may determine the mucosal dissemination pattern of MLP (Pals et al. 1994). MLP is regarded as a gastrointestinal variant of the mantle cell (MC) lymphoma and a major feature of MLP is the formation of multiple lymphomatous polyps along the gastrointestinal tract. Comparison of metastatic lymphoma cells derived from mucosal sites (MLP) and from lymph nodes (MC) revealed that α4β7 integrins are exclusively expressed on lymphoma cells from mucosal sites (Table 2). In contrast, expression of L-selectin, LFA-1, ICAM-1, or CD44 did not differ between MLP or MC lymphoma cells. These studies indicate that a tissue-specific homing mechanism mediated by α4β7 integrin and MAdCAM-1 may control the dissemination of MLP lymphoma cells to the intestinal mucosa (Pals et al. 1994). Analysis of α4β7 integrins on additional non-Hodgkin's lymphomas confirmed the crucial role of α4β7 integrins for tumor cell dissemination (Drillenburg et al. 1997). Thus, α4β7 integrins are expressed on T and B cell non-Hodgkin's lymphomas localized preferentially in mucosa-associated lymphoid tissues but are absent on non-Hodgkin's lymphomas derived from lymph nodes or cutaneous sites (Drillenburg et al. 1997).

6.2 Regulation of Tumor Cell Invasiveness by α4-Integrins

Homotypic cell adhesion of tumor cells is considered important for various steps of the metastatic cascade. Whereas the detachment of tumor cells from the primary tumor may be suppressed by homotypic adhesion, the aggregation of tumor cells and the formation of tumor cell emboli in circulation may facilitate the arrest of metastatic cells in the vasculature at distant sites (Reeves 1992; Weiss et al. 1988). Accordingly, a recent study has implicated α4β1 integrins in enhanced homotypic interactions and reduced metastasis formation of murine melanoma cells (Qian et al. 1994). In these experiments, three sublines of the murine B16 melanoma with different expression levels of α4β1 integrins were shown to markedly differ in their metastatic capacity after subcutaneous injection despite similar in vitro growth rates (Table 2). In contrast to intravenous injection of cancer cells, subcutaneous injection mimics the early steps of metastasis, beginning with cell growth at a primary tumor site. Under these conditions, B16 melanoma sublines with high expression of α4β1 showed decreased ability to form tumor colonies in the lung. Moreover, transfection of highly metastatic B16 melanoma cells with the α4 integrin subunit cDNA induced the low metastatic phenotype.

Immunohistochemical analysis of primary tumors derived from low metastatic B16 sublines revealed that α4 was expressed at high level on the tumor cell margins, whereas on primary tumors of highly metastatic melanoma cells expression of α4 integrins was nearly absent (Qian et al. 1994). When melanoma cells were injected intravenously, however, metastasis formation in lung was independent on α4 integrin expression. These findings suggest that suppression of the metastatic capacity may result from increased, α4 integrin-mediated homotypic tumor cell adhesion at the primary site. Consistent with this hypothesis, in vitro binding assays directly demonstrated homotypic adhesion of α4 integrin positive melanoma. Moreover, in vitro matrigel invasion of melanoma cells was reduced due to high level expression of α4 integrins. Treatment with an anti-α4 integrin antibody restored the ability to invade the substrates (Qian et al. 1994). These observations led to the conclusion that homophilic adhesion mediated by α4 integrins prevents the detachment of melanoma cells from the primary tumor thereby inhibiting invasiveness and subsequent metastasis formation.

In contrast to the site of primary tumor growth, homotypic adhesion of tumor cells in the circulation may facilitate the arrest of tumor cells in the microvasculature and enhance their metastatic capacity (LOTAN and RAZ 1983; URUSHIHARA et al. 1984; WEISS et al. 1988). However, QIAN et al. (1994) reported that melanoma cells overexpressing α4β1 integrins were found mostly as single cells in the lung vasculature indicating that homotypic adhesion of α4β1 integrin positive melanoma cells did not occur in circulation. These findings may explain why the expression of α4β1 integrins failed to enhance lung colonization after i.v. injection. However, it is conceivable that treatment of mice with cytokines that enhance the density of the α4 integrin ligand VCAM-1 at the endothelium may have increased lung metastasis formation similar to the results described above (GAROFALO et al. 1995; OKAHARA et al. 1994).

Another important mechanism controlling the invasive capacity of tumor cells is their ability to secrete proteolytic enzymes. A recent report indicates that α4β1 integrins are involved in the regulation of matrix metalloproteinase expression. Synovial fibroblasts, which express integrins α5β1 and α4β1, showed enhanced expression levels of certain metalloproteinases when plated on RGD-containing fibronectin fragments or anti-α5β1 integrin antibodies indicating that α5β1 integrin-mediated signals lead to the induction of matrix metalloproteinases (HUHTALA et al. 1995). Simultaneous addition of ligands that contain the fibronectin/CS-1 binding motif for α4β1 integrins or complete fibronectin that contains both the RGD site and the CS-1 fragment, however, suppressed the induction of metalloproteinases (HUHTALA et al. 1995). It therefore appears that α4β1 integrin mediated signals may inhibit the induction of metalloproteinases in a dominant manner. Thus, it is conceivable that ligand binding to α4β1 integrin also down-regulates matrix metalloprotease secretion by melanoma cells leading to reduced tissue invasion and impaired metastasis formation. Regulation of matrix metalloproteinases could therefore provide an alternative explanation for the reduced metastasis formation of α4β1-integrin-positive B16 melanoma cells after subcutaneous injection (QIAN et al. 1994).

Together, the various results from experimental tumor models using melanoma or sarcoma cells suggest that the effects mediated by α4 integrins differ dependent on the stage of tumor progression at which α4 integrins are induced. While induction of α4 integrin expression on primary tumors may support homotypic tumor cell interactions or inhibit metalloproteinase secretion resulting in reduced metastasis formation, increased expression of α4 integrins on tumor cells that have entered blood circulation may promote interactions with VCAM-1 positive endothelium thereby enhancing the frequency of metastasis formation.

6.3 Expansion of Disseminated Lymphoid Tumor Cells is Inhibited by α4-Integrins

Using a murine lymphoma model, we have recently shown that α4 integrins may mediate tumor suppressive effects (GOSSLAR et al. 1996). Cell lines from the murine T cell lymphoma LB were established that differ exclusively in the expression of α4 integrins. Despite similar growth rates in vitro significantly different metastatic capacities were observed after i.v. injection into syngenic mice. It was demonstrated that metastasis formation of LB-α4 cells in a large number of lymphoid or non-lymphoid organs including spleen, lymph nodes, Peyer's patches, lung, liver, and kidney was greatly reduced when compared with LB-NTK control cells. In marked contrast, expansion of metastatic lymphoma cells in bone marrow was not affected by the expression of α4 integrins suggesting that bone marrow may represent a unique compartment for the regulation of metastasis formation by α4 integrins. Interestingly, in a recent report it was shown that de novo expression of α4 integrin in CHO cells promotes tumor formation in bone marrow of nude mice (MATSUURA et al. 1996).

In vivo homing experiments with ^{51}Cr-labeled lymphoma cells revealed that LB-α4 and LB-NTK cells accumulated at least with same efficiency in lymphoid and nonlymphoid organs. In mucosal lymphoid organs such as Peyer's patches and mesenteric lymph nodes the accumulation of LB-α4 cells was selectively enhanced compared to control LB-NTK cells. Histopathological analysis revealed that at early time points of metastatic growth both LB-α4 and LB-NTK cells were predominantly localized to the T cell areas of lymphoid organs. These results therefore confirmed the intact migratory capacity of LB-α4 cells. In addition, the ^{51}Cr labeling experiments revealed that the number of circulating or extravasated LB-α4 and LB NTK cells were comparable for up to 48 h after i.v. injection into mice. These observations argue against an enhanced susceptibility of LB-α4 cells to NK (natural killer) cell lysis in vivo, because NK cell mediated tumor cell clearance was shown to be rapid and to occur within 24 h after i.v. injection of tumor cells (HANNA and FIDLER 1980, 1981). Collectively, these data strongly suggest that the inhibitory effect of α4 integrins on lymphoma metastasis formation occurs at a stage subsequent to the infiltration of target organs (GOSSLAR et al. 1996).

The studies using the LB lymphoma model identify additional events of the metastatic cascade that may be regulated by α4 integrins. Consistent with previous findings (DRILLENBURG et al. 1997; GAROFALO et al. 1995; OKAHARA et al. 1994; PALS et al. 1994), the in vivo migration experiments using lymphoma cells support the concept that expression of α4 integrins promotes accumulation of tumor cells in selected organs (see above). For some tumors, e.g., melanoma cells (GAROFALO et al. 1995; OKAHARA et al. 1994), this mechanism may result in enhanced metastasis formation. In contrast to these models, however, the formation of lymphoma cell metastases may be impaired by α4 integrin dependent events. These inhibitory mechanisms operate at a stage subsequent to the invasion of target organs and appear to dominate over positive effects on tumor cell accumulation. Therefore, in some tumors proliferation or survival of metastatic cells may also depend on α4 integrins.

A possible mechanism by which α4 integrins may influence the growth and expansion of lymphoma cells is via adhesion triggered signal transduction events. The differentiation and survival of T and B lymphocytes in lymphatic organs was shown to depend on α4 integrin mediated signals (Burkly et al. 1991; Damle et al. 1993; Koopman et al. 1994). In general, signals may be transmitted either by cytoplasmic proteins linked to α4 integrins or by associated transmembrane receptors thereby generating unique cell surface-bound signaling complexes. Recently it was reported that α4β1 and α4β7 integrins associate with several members of the transmembrane-4 superfamily (TM4SF) including CD9, CD63, CD81, and CD82 (Dong et al. 1995; Ikeyama et al. 1993; Mannion et al. 1996; Radford et al. 1995). Interestingly, CD81 (TAPA-1) has previously been described as the "target of an antiproliferative antibody." It was shown that antibody cross-linking of CD81 inhibits the proliferation of B lymphoma cells in vitro (Oren et al. 1990).

Consistent with a role of TM4SF proteins in tumor progression, B16 melanoma cells transfected with cDNA encoding CD9 were shown to form fewer pulmonary metastases after i.v. injection than controls. Further in vitro studies suggested that the reduced metastatic capacity may result from impaired cell motility (Ikeyama et al. 1993). Prostate carcinoma cells transfected with CD82, another TM4SF member, showed reduced metastatic capacity after s.c. injection into nude mice, whereas the cell growth at the primary tumor site, i.e., the tumorigenicity, was not altered (Dong et al. 1995). Unlike tumor cells transfected with CD9 or CD82, expression of CD63 in human melanoma cells resulted in a reduced tumorigenicity after s.c. injection into nude mice suggesting a suppressive effect of CD63 on melanoma cell growth in vivo (Radford et al. 1995). Taken together, overexpression of TM4SF proteins appears to affect tumor formation by various mechanisms. So far it remains unclear, however, whether association with integrins is required to induce TM4SF dependent anti-tumor effects including reduced growth and motility. It is, however, tempting to speculate that ligation of TM4SF associated integrins may induce TM4SF dependent signaling resulting in tumor suppressive effects that are triggered by integrins, but mediated by TM4SF proteins.

Recently, we have demonstrated that the TM4SF protein CD81 (TAPA-1), but not CD9, is expressed on LB lymphoma cells (M. Bittner and B.Holzmann, unpublished observations). It is therefore conceivable that ligand induced cross-linking of α4 integrins may result in coclustering of CD81 on lymphoma cells. As a consequence, TM4SF dependent signaling pathways may be activated that mediate impaired proliferation or survival of disseminated lymphoma cells. According to this hypothesis, the biological response of tumor cells to α4 integrin expression and ligand induced ligation may be critically controlled by the expression pattern and surface density of regulatory coreceptors.

7 Summary

Taken together, α4 integrins may influence metastatic process at various stages (Fig. 1). The detachment of tumor cells from the primary tumor and the invasion of the surrounding tissue represent the onset of tumor metastasis. There is good experimental evidence that at the primary tumor site expression of α4 integrins inhibits the ability of melanoma cells to break loose. This could be achieved either by strengthening of homotypic adhesion to adjacent tumor cells or by down regulation of matrix metalloproteases that are required for tumor cell migration through the extracellular matrix. After entering the blood circulation, α4 integrins on tumor cells derived from melanomas, sarcomas or lymphomas rather promote than inhibit accumulation of disseminated cells in distant organs. The positive effects of α4 integrins at this stage of metastasis formation appear to depend on α4 integrin interactions with ligands expressed on the surface of endothelial cells. While VCAM-1 is expressed on endothelial cells exposed to inflammatory cytokines, MAdCAM-1 is constitutively expressed on mucosal endothelium. In addition, it is conceivable that tumor cell aggregates trapped in the microcirculation may trigger local inflammatory reactions that result in VCAM-1 up-regulation. Tumor cell-bound α4 integrins may strengthen adhesion to endothelium and promote trans-endothelial migration (Hauzenberger et al. 1997; Meerschaert and Furie 1994).

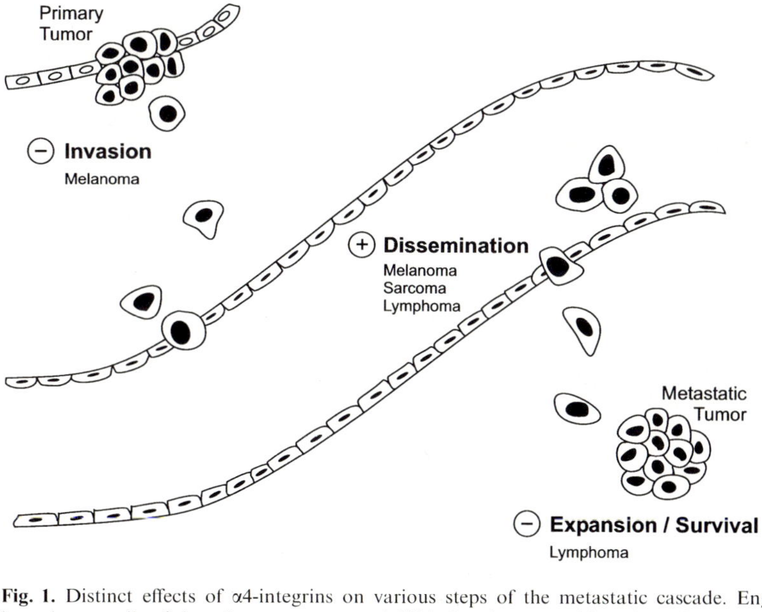

Fig. 1. Distinct effects of α4-integrins on various steps of the metastatic cascade. Engagement of α4 integrins on cells of the primary tumor may inhibit detachment and invasion of malignant cells, whereas expression on circulation tumor cells may enhance dissemination. In some tumors, expansion or survival of tumor cells that have lodged to secondary sites is impaired by α4 integrins. Tumor models for which these various effects have been demonstrated are indicated

Successful formation of new tumor colonies in distant organs is the final step in the metastatic cascade. Interestingly, α4 integrin dependent mechanisms may either promote or inhibit this process. Thus, it was observed that α4 integrins may direct cancer cells like CHO and lymphoma cells to organ compartments, where ligands for α4 integrins are expressed (e.g., bone marrow). Depending on the tumor type this event may result in enhanced metastasis formation. However, as was documented for murine lymphoma cells α4 integrins may also inhibit tumor cell growth either by inducing apoptosis or by reducing the proliferation rate.

Based on numerous studies on human cancers and experimental tumor models, α4 integrins may represent attractive target molecules for therapeutic manipulation of tumor cell behavior. To this end, however, it will be of great importance to precisely define the molecular basis for the adverse effects of α4 integrins on metastasis formation.

Acknowledgement. This work was supported by a grant from the Deutsche Krebshilfe, Dr. Mildred Scheel Stiftung.

References

Aldelba SM, Mette SA, Elder DE, Stewart R, Damjanovich L, Herlyn M, Buck CA (1990)-integrin distribution in malignant melanoma: association of the β3 subunit with tumor progression. Cancer Res 50:6757–6764

Altevogt P, Hubbe M, Ruppert M, Lohr J, von Hoegen P, Sammar M, Andrew DP, McEvoy L, Humphries MJ, Butcher EC (1995) The α4-integrin chain is a ligand for α4β7 and α4β1. J Exp Med 182:345–355

Bell RG, Issekutz T (1993) Expression of a protective intestinal immune response can be inhibited at three distinct sites by treatment with anti-α4-integrin. J Immunol 151:4790–4802

Berlin C, Berg EL, Briskin MJ, Andrew DP, Kilshaw PJ, Holzmann B, Weissman IL, Hamann A, Butcher EC (1993) α4β7-integrin mediates lymphocyte binding to the mucosal vascular addressin MAdCAM-1. Cell 74:185–195

Briskin MJ, McEvoy LM, Butcher EC (1993) MAdCAM-1 has homology to immunoglobulin and mucin-like adhesion receptors and to IgA1. Science 363:461–464

Briskin MJ, Rott L, Butcher EC (1996) Structural requirements for mucosal vascular addressin binding to its lymphocyte receptor α4β7: common themes among-integrin-Ig family interactions. J Immunol 156:719–726

Burkly LC, Jakubowski A, Newman BM, Rosa MD, Chi-Rosso G, Lobb RR (1991) Signaling by vascular cell adhesion molecule-1 (VCAM-1) through VLA-4 promotes CD3-dependent T cell proliferation. Eur J Immunol 21:2871–2875

Carlos TM, Harlan JM (1994) Leucocyte-endothelial adhesion molecules. Blood 7:2068–2101

Damle NK, Aruffo A (1991) Vascular cell adhesion molecule-1 induces T-cell antigen receptor-dependent activation of CD4+ T lymphocytes. Proc Natl Acad Sci USA 88:6403–6407

Damle NK, Klussman K, Leytze G, Aruffo A, Linsley PS, Ledbetter JA (1993) Costimulation with-integrin ligands intercellular adhesion molecule-1 or vascular cell adhesion molecule-1 augments activation-induced death of antigen-specific CD4+ T-lymphocytes. J Immunol 151:2368–2379

Damle NK, Klussman K, Leytze G, Myrdal S, Aruffo A, Ledbetter JA, Linsley PS (1994) Costimulation of T lymphocytes with-integrin ligands intercellular adhesion molecule-1 or vascular cell adhesion molecule-1 induces functional expression of CTLA-4, a second receptor for B7. J Immunol 152 (6):2686–2697

Dong JT, Lamb PW, Rinker-Schaeffer CW, Vukanovic J, Ichikawa T, Isaacs JT, Barrett JC (1995) KAI1, a metastasis suppressor gene of prostate cancer on human chromosome 11p11.2. Science 268:884–886

Drillenburg P, van der Voort R, Koopman G, Dragosics B, van Krieken JHJM, Kluin P, Meenan J, Lazarovits AI, Radaszkiewicz T, Pals ST (1997) Preferential expression of the mucosal homing receptor-integrin α4β7 in gastrointestinal non-Hodgkin's lymphomas. Am J Pathol 150(3):919–927

Elices MJ, Tsai V, Strahl D, Goel AS, Tollefson V, Arrhenius T, Wayner EA, Gaeta FCA, Fikes JD, Firestein GS (1994) Expression and functional significance of alternatively spliced CS1 fibronectin in rheumatoid arthritis microvasculature. J Clin Invest 93:405–416

Ennis E, Isberg RR, Shimizu Y (1993) Very late antigen 4-dependent adhesion and costimulation of resting human T cells by the bacterial β1-integrin ligand invasin. J Exp Med 177:207–212

Erle DJ, Briskin MJ, Butcher EC, Garcia-Pardo A, Lazarovits AI, Tidswell M (1994) Expression and function of the MAdCAM-1 receptor, -integrin α4β7, on human leukocytes. J Immunol 153:517–528

Freedman AS, Munro JM, Rice GE, Bevilacqua MP, Morimoto C, McIntyre BW, Rhynhart K, Pober JS, Nadler LM (1990) Adhesion of human B cells to germinal centers in vitro involves VLA-4 and INCAM-110. Science 249:1030–1033

Garofalo A, Chirivi RGS, Foglieni C, Pigott R, Mortarini R, Martin-Padura I, Anichini A, Gearing AJ, Sanchez-Madrid F, Dejana E, Giavazzi R (1995) Involvement of the very late antigen 4-integrin on melanoma in interleukin-1-augmented experimental metastasis. Cancer Res 55:414–419

Gosslar U, Jonas P, Luz A, Lifka A, Naor D, Hamann A, Holzmann B (1996) Predominant role of α4-integrins for distinct steps of lymphoma metastasis. Proc Natl Acad Sci USA 93:4821–4826

Hanna N, Fidler IJ (1980) Role of natural killer cells in the destruction of circulating tumor emboli. J Natl Cancer Inst 65(4):801–809

Hanna N, Fidler IJ (1981) Relationship between metastatic potential and resistance to natural killer cell-mediated cytotoxicity in three murine tumor systems. J Natl Cancer Inst 66(6):1183–1190

Hauzenberger D, Klominek J, Holgersson J, Bergstrom SE, Sundqvist KG (1997) Triggering of motile behavior in T lymphocytes via cross-linking of α4β1 and αLβ2. J Immunol 158(1):76–84

Hemler ME, Elices MJ, Parker C, Takada Y (1990) Structure of the-integrin VLA-4 and its cell-cell and cell-matrix adhesion functions. Immunol Rev 114:45–65

Holzmann B, Weissman IL (1989) Peyer's patch-specific lymphocyte homing receptors consist of a VLA-4-like a chain associated with either of two inegrin β chains, one of which is novel. EMBO J 8 (6):1735–1741

Holzmann B, McIntyre BW, Weissman IL (1989) Identifaction of a murine Peyer's patch-specific lymphocyte homing receptor as an-integrin molecule with a chain homologous to human VLA-4α. Cell 56:37–46

Hu MC-T, Crowe DT, Weissman IL, Holzmann B (1992) Cloning and expression of mouse-integrin bp(β7): a functional role in Peyer's patch-specific lymphocyte homing. Proc Natl Acad Sci USA 89:8254–8258

Huhtala P, Humphries MJ, McCarthy JB, Tremble PM, Werb Z, Damsky CH (1995) Cooperative signaling by α5β1 and α4β1-integrins regulates metalloproteinase gene expression in fibroblasts adhering to fibronectin. J Cell Biol 129(3):867–879

Hynes RO (1992)-integrins: versatility, modulation, and signalling in cell adhesion. Cell 69:11–25

Ikeyama S, Koyama M, Yamaoko M, Sasada R, Miyake M (1993) Suppression of cell motility and metastasis by transfection with human motility-related protein (MRP-1/CD9) DNA. J Exp Med 177:1231–1237

Issekutz TB, Miyasaka M, Issekutz AC (1996) Rat blood neutrophils express very late antigen 4 and it mediates migration to arthritic joint and dermal inflammation. J Exp Med 183:2175–2184

Kilger G, Holzman B (1995) Molecular analysis of the physiological and pathophysiological role of α4-integrins. J Mol Med 73:347–354

Kilger G, Clements J, Holzmann B (1997) Amino acid residues required for binding of vascular cell adhesion molecule-1 to-integrin α4β1. Int Immunol 9 (2):219–226

Kilshaw PJ, Murant SJ (1990) A new surface antigen on intraepithelial lymphocytes in the intestine. Eur J Immunol 20:2201–2207

Komoriya A, Green LJ, Mervic M, Yamada SS, Yamada KM, Humphries MJ (1991) The minimal essential sequence for a major cell type-specific adhesion site (CS1) within the alternatively spliced type III connecting segment domain of fibronectin is leucine-aspartic acid-valine. J Biol Chem 266 (23):15075–15079

Koopman G, Parmentier HK, Schuurman H-J, Newman W, Meijer CJLM, Pals ST (1991) Adhesion of human B cells to follicular dendritic cells involves both the lymphocyte function-associated antigen 1/intercellular adhesion molecule 1 and very late antigen 4/vascular cell adhesion molecule 1 pathways. J Exp Med 173:1297–1304

Koopman G, Keehnen RMJ, Lindhout E, Newman W, Shimizu Y, van Seventer GA, de Groot C, Pals ST (1994) Adhesion through the LFA-1 (CD11a/CD18) – ICAM-1 (CD54) and the VLA-4 (CD49d)-VCAM-1 (CD106) pathways prevents apoptosis of germinal center B cells. J Immunol 152:3760–3766

Levesque J-P, Leavesley DI, Niutta S, Vadas M, Simmons PJ (1995) Cytokines increase human hemopoietic cell adhesiveness by activation of very late antigen (VLA)-4 and VLA-5-integrins. J Exp Med 181:1805–1815

Liotta LS, Steeg PS, Stetler-Stevenson WG (1991) Cancer metastasis and anigiogenesis: an imbalance of positive and negative regulation. Cell 64:327–336

Lloyd AR, Oppenheim JJ, Kelvin DJ, Taub DD (1996) Chemokines regulate T cell adherence to recombinant adhesion molecules and extracellular matrix proteins. J Immunol 156:932–938

Lotan R, Raz A (1983) Low colony formation in vivo and in culture as exhibited by metastatic melanoma cells selected for reduced homotypic aggregation. Cancer Res 43:2088–2093

Mannion BA, Berditchevski F, Kraeft S-K, Chen LB, Hemler ME (1996) Transmembrane-4 superfamily proteins CD81 (TAPA-1) CD82, CD63 and CD53 specifically associate with-integrin α4β1 (CD49d/CD29). J Immunol 157:2039–2047

Masinovsky B, Urdal D, Gallatin WM (1990) IL-4 acts synergistically with IL-1β to promote lymphocyte adhesion to microvascular endothelium by induction of vascular cell adhesion molecule-1. J Immunol 145(9):2886–2895

Matsuura N, Puzon-McLaughlin W, Irie A, Morikawa Y, Kakudo K, Takada Y (1996) Induction of experimental bone metastasis in mice by transfection of-integrin α4β1 into tumor cells. Am J Pathol 148(1):55–61

Mattila P, Majuri M-L, Renkonen R (1992) VLA-4-integrin on sarcoma cell lines recognizes endothelial VCAM-1. Differential regulation of the VLA-4 avidity on various sarcoma cell lines. Int J Cancer 52:918–923

Meerschaert J, Furie MB (1994) Monocytes use either CD11/CD18 or VLA-4 to migrate across human endothelium in vitro. J Immunol 152(4):1915–1926

Mosher DF (1990) Physiology of thrombospondin. Annu Rev Med 41:85–97

Mould AP, Wheldon LA, Komoriya A, Wayner EA, Yamada KM, Humphries MJ (1990) Affinity chromatographic isolation of the melanoma adhesion receptor for the IIICS region of fibronectin and its identification as the-integrin α4β1. J Biol Chem 265(7):4020–4024

Mould AP, Komoriya A, Yamada KM, Humphries MJ (1991) The CS5 peptide is a second site in the IIICS region of fibronectin recognized by the-integrin α4β1. Inhibition of α4β1 function by RGD peptide homologues. J Biol Chem 266(6):3579–3585

Nicolson GL (1982) Cancer metastasis: organ colonization and the cell-surface properties of malignant cells. Biochim Biophys Acta 695:113–176

Nojima Y, Humphries MJ, Mould AP, Komoriya A, Yamada KM, Schlossman SF, Morimoto C (1990) VLA-4 mediates CD3-dependent CD4+ T cell activation via the CS1 alternatively spliced domain of fibronectin. J Exp Med 172(4):1185–1192

Okahara H, Yagita H, Miyake K, Okumura K (1994) Involvement of very late antigen 4 (VLA-4) and vascular cell adhesion molecule 1 (VCAM-1) in tumor necrosis factor a enhancement of experimental metastasis. Cancer Res 54:3233–3236

Oren R, Takahashi S, Doss C, Levy R, Levy S (1990) TAPA-1, the target of an antiproliferative antibody, defines a new family of transmembrane proteins. Moll Cell Biol 10(8):4007–4015

Osborn L, Vassallo C, Browning BG, Tizard R, Haskard DO, Benjamin CD, Dougas I, Kirchhausen T (1994) Arrangement of domains, and amino acid residues required for binding of vascular cell adhesion molecule-1 to its counter-receptor VLA-4 (α4β1). J Cell Biol 124(4):601–608

Paavonen T, Thsala S, Majuri M-L, Böhling T, Renkonen R (1994) In vivo evidence of the role of α4β1-VCAM-1 interaction in sarcoma, but not in carcinoma extravasation. Int J Cancer 58:298–302

Pals ST, Drillenburg P, Dragosics B, Lazarovits AI, Radaszkiewicz T (1994) Expression of the mucosal homing receptor α4β7 in malignant lymphomatous polyposis of the intestine. Gastroenterology 107:1519–1523

Postigo AA, Sanchez-Mateos P, Lazarovits AI, Sanchez-Madrid F, de Landazuri MO (1993) α4β7-integrin mediates B cell binding to fibronectin and vascular cell adhesion molecule-1. J Immunol 151:2471–2483

Qian F, Vaux DL, Weissman IL (1994) Expression of the-integrin α4β7 on melanoma cells can inhibit the invasive stage of metastasis formation. Cell 77:335–347

Radford KJ, Mallesch J, Hersey P (1995) Suppression of human melanoma cell growth and metastasis by the melanoma-associated antigen CD63 (Me491). Int J Cancer 62:631–635

Reeves ME (1992) A metastatic tumor cell line has greatly reduced levels of a specific homotypic cell adhesion molecule activity. Cancer Res 52:1546–1522

Rincon J, Prieto J, Patarroyo M (1992) Expression of-integrins and other adhesion molecules in Epstein-Barr Virus transformed B lymphoblastoid cells and Burkitt's lymphoma cells. Int J Cancer 51: 452–458

Ruegg C, Postigo AA, Sikorski EE, Butcher EC, Pytela R, Erle DJ (1992) Role of-integrin α4β7/α4βp in lymphocyte adherence to fibronectin and VCAM-1 and in homotypic cell clustering. J Cell Biol 117(1):179–189

Sanchez-Aparicio P, Dominguez-Jimenez C, Garcia-Pardo A (1994) Activation of the α4β1-integrin through the β1 subunit induces recognition of the RGDS sequence in fibronectin. J Cell Biol 126(1):271–279

Schweighofer T, Tanaka Y, Tidswell M, Erle DJ, Horgan KJ, Ginther LGE, Lazarovits AI, Buck D, Shaw S (1993) Selective expression of-integrin α4β7 on a subset of human CD4+ memory T cells with hallmarks of gut-tropism. J Immunol 151:517–528

Shimizu Y, van Seventer GA, Horgan KJ, Shaw S (1990) Regulated expression and binding of three VLA (β1)-integrin receptors on T cells. Science 345:250–253

Stetler-Stevenson WG, Aznavoorian S, Liotta LA (1993) Tumor cell interactions with the extracellular matrix during invasion and metastasis. Annu Rev Cell Biol 9:541–573

Strauch UG, Lifka A, Goßlar U, Kilshaw PJ, Clements J, Holzmann B (1994) Distinct binding specificities of-integrins α4β7 (LPAM-1), α4β1 (VLA-4), and αIELβ7. Int Immunol 6(2):263–275

Streeter PR, Berg EL, Rouse BTN, Bargatze RF, Butcher EC (1988) A tissue-specific endothelial cell molecule involved in lymphocyte homing. Science 331:41–46

Taichman DB, Cybulsky MI, Djaffar I, Longenecker BM, Teixido J, Rice GE, Aruffo A, Bevilacqua MP (1991) Tumor cell surface α4β1-integrin mediates adhesion to vascular endothelium: demonstration of an interaction with the N-terminal domains of INCAM-110/VCAM-1. Cell Regul 2:347–355

Thornhill MH, Haskard DO (1990) IL-4 regulates endothelial cell activation by IL-1, tumor necrosis factor, of IFN-γ. J Immunol 145:865–872

Tiisala S, Paavonen T, Renkonen R (1995) αEβ7 and α4β7-integrins associated with intraepithelial and mucosal homing, are expressed on macrophages. Eur J Immunol 25(2):411–417

Udagawa T, Woodside DG, McIntyre BW (1996) α4β1 (CD49d/CD29)-integrin costimulation of human T cells enhances transcription factor and cytokine induction in the absence of altered sensitivity to anti-CD3 stimulation. J Immunol 157:1965–1972

Urushihara H, Ikawa Y, Tsuruo T (1984) Adhesive properties of weakly and highly metastatic melanoma cell lines. Gann 75:534–539

van Seventer GA, Newman W, Shimizu Y, Nutman TB, Tanaka Y, Horgan KJ, Gopal TV, Ennis E, O'Sullivan D, Grey H, Shaw S (1991) Analysis of T cell stimulation by superantigen plus major histocompatibility complex class II molecules or by CD3 monoclonal antibody: costimulation by purified adhesion ligands VCAM-1, ICAM-1 but not ELAM-1. J Exp Med 174:901–913

Viney JL, Jones S, Chiu HH, Lagrimas B, Renz ME, Presta LG, Jackson P, Hillman KJ, Lew S, Fong S (1996) Mucosal addressin cell adhesion molecule-1: a structural and functional analysis demarcates the-integrin binding motif. J Immunol 157(6):2488–2497

Vonderheide RH, Springer TA (1992) Lymphocyte adhesion through very late antigen 4: evidence for a novel binding site in the alternatively spliced domain of vascular cell adhesion molecule 1 and an additional α4-integrin counter-receptor on stimulated endothelium. J Exp Med 175(6):1433–1442

Weiss L, Orr FW, Honn KV (1988) Interaction of cancer cells with the microvasculature during metastasis. FASEB J 2:12–21

Wellicome SM, Thornhill MH, Pitzalis C, Thomas DS, Lanchbury JSS, Panayi GS, Haskard DO (1990) A monoclonal antibody that detects a novel antigen on endothelial cells that is induced by tumor necrosis factor, IL-1, or lipopolysaccharide. J Immunol 144(7):2558–2565

Yabkowitz R, Dixit VM, Guo N, Roberts DD, Shimizu Y (1993) Activated T-cell adhesion to thrombospondin is mediated by the α4β1 (VLA-4) and α5β1 (VLA-5)-integrins. J Immunol 151(1):149–158

Zhang Z, Vuori K, Wang H, Reed JC, Rouslahti E (1996) α_4-integrin activation by R-ras. Cell 85:61–69

Organ-Specific Requirements for Cell Adhesion Molecules During Lymphoma Cell Dissemination

D. Naor[1], R. Vogt Sionov[1], M. Zahalka[1], M. Rochman[1], B. Holzmann,[2] and D. Ish-Shalom[1]

1 Introduction . . . 143
2 LB Cells Use Different Migration Routes for Spleen and Lymph Node Invasion . . . 148
3 LB Cells Use Distinct Adhesion Molecules for Spleen and Lymph Node Invasion . . . 152
4 Different Strategies Exercised by LB Cells for Their Establishment in Spleen and Lymph Nodes . . . 155
5 The LB Cell CD44 Ligand . . . 157
6 Conclusions . . . 161
References . . . 162

1 Introduction

The malignant process is, in many aspects, a distorted image of normal physiological activities. In this respect, the dissemination mechanism of malignant lymphomas may display a hazy reflection of normal cell migration, an essential function of the lymphoid system. The successful response of the individual's defense machinery against invading microorganisms is largely due to its ability to rapidly mobilize leukocytes to the site of infection. Cell motility in blood, lymph, lymphoid organs and tissues is highly dependent on the coordinated activity of different cell adhesion molecules. These are implicated in the transendothelial migration of intravasated and extravasated cells, the capture of cells by the luminal surface of the endothelium, cell rolling and cell arrest in the vasculature, binding of lymphocytes to the high endothelial venule (HEV) of the lymph node, as well as subsequent cell lodgment in organ parenchyma (a process known as cell homing), and cell migration on extracellular matrix (ECM) (Parkhurst and Saltzman 1992; Picker and Butcher 1992; Thomas et al. 1992; Springer 1994; Ley and Tedder 1995). Three pairs of adhesion receptor and counterreceptor families are implicated in the interaction between leukocytes and their target cells in the endothelium and tissues: (1) integrins, which target molecules of the immunoglobulin superfamily (IgSF) or ECM components, (2) se-

[1]The Lautenberg Center for General and Tumor Immunology, The Hebrew-University-Hadassah Medical School, POB 12272, Jerusalem 91120, Israel
[2]Institute for Medical Microbiology and Hygiene, Technical University, 81675, Munich, Germany

lectins, which interact with sialylated carbohydrate determinants O-linked to mucin-like molecules (also known as addressins), and (3) CD44 receptors with binding affinity for matrix and cell surface constituents (Ruoslahti 1991; Yamada 1991; Picker and Butcher 1992; Lesley et al. 1993; Springer 1994; Naor et al. 1997). It is generally agreed that the reciprocal interaction between endothelial cell P- and E-selectins and leukocyte L-selectin with the corresponding leukocyte and endothelial cell addressins mediates the initial capture of leukocytes from the flowing blood and their roll along the blood vessels. Cell rolling is halted by a firm interaction between the leukocyte integrins (Mac-1, lymphocyte function-associated antigen [LFA-1], very late antigen-4 [VLA-4]) and endothelial-cell IgSF target molecules (intercellular adhesion molecule-1 [ICAM-1], ICAM-2 and vascular cell adhesion molecule [VCAM-1]). The transition from the rolling phase to firm attachment is activated by cytokines or chemokines accumulating at the inflammation site. Following the arrest stage, the leukocytes initiate a process of transendothelial migration, which is terminated by their localization in the inflamed tissue. This process is also mediated by leukocyte integrins and their endothelial cell IgSF counterparts. The selectin phase of leukocyte migration to the inflamed site overlaps, to a certain extent, with the integrin phase, as selectins participate in the leukocyte arrest and diapedesis, whereas integrins play a role in the cell rolling stage. Cross-talk between selectins, chemokine receptors and integrins orchestrates the entire leukocyte migration process, enabling a rapid and efficient response to infection by microorganisms (Butcher 1991; Picker and Butcher 1992; Springer 1994).

CD44 is another cell surface molecule that influences cell migration and cell lodgment in lymphoid organs or inflamed tissues. The molecule is a single chain glycoprotein comprising a conserved NH_2-terminal domain, a nonconserved membrane proximal region, a conserved transmembrane-spanning domain and a conserved cytoplasmic tail that can interact with the cytoskeleton. The genomic sequence of CD44 includes five constant exons at the 5′ terminal, five constant exons at the 3′ end, and ten variant exons in the middle (designated V1, V2, V3....V10). Differential alternative splicing generates a variable region in the CD44 transcript (generally designated CD44v), which contains different combinations of variant exons. Insertion of exons V4, V5, V6 and V7 generates the pMeta-1 CD44 which, upon transfection, confers metastatic potential on nonmetastatic rat pancreatic adenocarcinoma cells (Günthert et al. 1991). Insertion of variant exons V8, V9 and V10 gives rise to epithelial cell CD44. Insertion of variant exons V3–V10 in tandem produces keratinocyte CD44, one of the longest CD44 isoforms known. Standard CD44 (CD44s), which lacks the entire variable region and is expressed preferentially on hematopoietic cells, is also known as CD44H (Fig. 1). The ligand binding site of CD44 is included in the NH_2-terminal extracellular domain. The principal ligand of CD44 is hyaluranic acid (HA; hyaluronate, hyaluronan), but other matrix components (collagen, fibronectin, laminin and chondroitin sulfate), as well as nonmatrix constituents (mucosal vascular addressin, serglycin, osteopontin and the class II invariant chain), can interact with this receptor. The structural polymorphism of the CD44 molecule may explain its multifunctional nature and its ability to interact with many ligands. The CD44 glycoprotein is involved in cell-cell and cell-matrix interactions, as well as in cell traffic on endothelium or ECM components. The molecule collaborates (in humans but not in mice) with selectins and integrins in the process of lymph node homing by binding the lymphocytes to HEV. In addition, CD44 is implicated in the presentation of cytokines, chemokines and growth factors to traveling cells. The molecule is also involved in the transmission of growth signals, as well as signals mediating hematopoiesis and prevention of apoptosis. Finally, uptake and intracellular degradation of hyaluronic acid can be mediated by the CD44 receptor. Marked accumulation of CD44, and in some cases also of hyaluronan, is detected in areas of intensive cell migration and cell proliferation, as in wound healing, tissue remodeling, inflammation, morphogenesis and carcinogenesis (reviewed in Lesley et al. 1993; Naor et al. 1997).

It has been demonstrated in certain experimental models that extravasation of lymphocytes and their subsequent localization in different organs is dependent on a specific interaction between the lymphocyte adhesion receptor and the target cell's counterreceptor. These cell surface molecules are used as "key and lock codes," enabling cells expressing different adhesion phenotypes to lodge in distinct lymphoid organs or inflamed sites. The following are a few examples: Homing to peripheral lymph nodes is targeted by lymphocyte L-selectin to peripheral lymph node vascular addressin (PNAd) of the organ HEV. This specific pairing is stabilized by the association between the activated integrin LFA-1 ($\alpha L\beta 2$) and molecules of the ICAM family (Picker and Butcher 1992; Springer 1994). The interaction between lymphocyte CD44 and an unknown ligand of human HEV may further enhance the peripheral lymph node homing process (Jalkanen et al. 1987; Pals et al. 1989; Toyama-Sorimachi et al. 1993). Lymphocyte homing to Peyer's patch is mediated by a different set of adhesion molecules: lymphocyte integrin $\alpha 4\beta 7$ and HEV mucosal vascular addressin (MAd). It has not been resolved whether these molecules are indeed directly paired, as suggested by Hamann et al. (1994), or if each of them targets a different counterreceptor. Adhesion molecules LFA-1 and CD44 may further increase binding of the lymphocytes to Peyer's patch HEV, facilitating their homing. Infiltration of memory T cells into the skin is mediated by a specific interaction between the addressin cutaneous lymphocyte-associated antigen (CLA) of the lymphocyte and the endothelial-leukocyte adhesion molecule-1 (ELAM-1, E-selectin of skin vasculature). The specific pairing is presumably strengthened by the interaction between lymphocyte VLA-4 and cutaneous endothelium VCAM-1 (Picker and Butcher 1992; Springer 1994). These interactions explain why lymphocytes removed from one of the above mentioned sites preferentially recirculate, following their injection, to the organ from which they were collected (Mackay 1992). It is not clear whether a subpopulation of cells a priori expresses organ- or site-specific adhesion molecules, or if local factors induce the cells to express the specific receptor after their random lodgement in the target tissue. Selective up-regulation of a particular set of surface molecules following their interaction with the tissue counterreceptor is another possible mechanism for generating the adhesive key and lock pairing.

Disseminating neoplastic cells may exhibit the same or an altered version of the chemotactic activity and adhesion-dependent homing function displayed by normal cells for circulation and localization in target tissues. Organ-derived chemotactic fac-

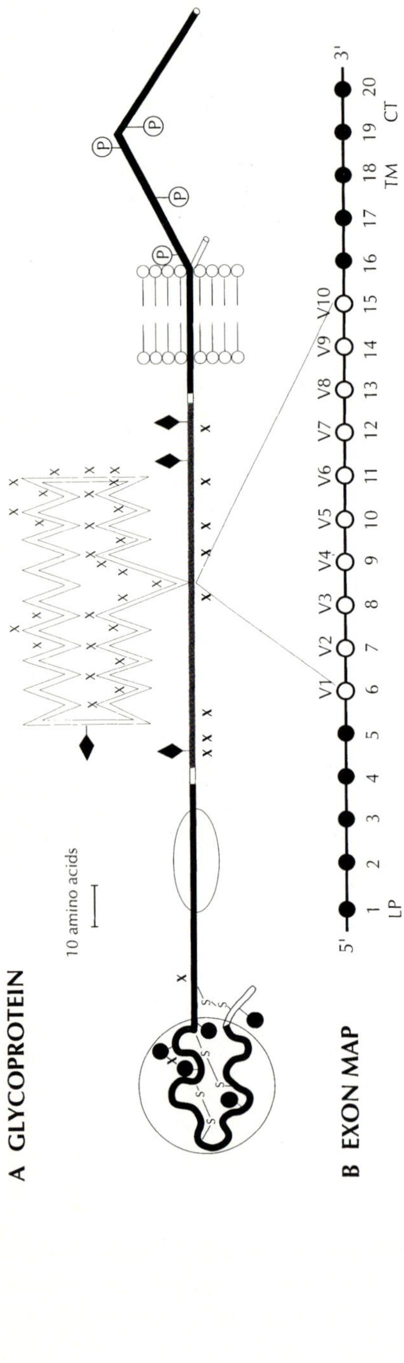
A GLYCOPROTEIN
10 amino acids
P
B EXON MAP
5'
3'
1 2 3 4 5 6 7 8 9 10 11 12 13 14 15 16 17 18 19 20
LP
V1 V2 V3 V4 V5 V6 V7 V8 V9 V10
TM
CT
C ALTERNATIVELY SPLICED TRANSCRIPTS
1. CD44s (short tail)
2. CD44s (long tail)
3. pMeta-1
4. pMeta-2
5. Epithelial CD44
6. CD44v of keratinocytes

tors, present in soluble form or confined to the cell membrane or ECM, are involved in the selective navigation of metastatic cells and in the dissemination of lymphoma cells to specific sites (YEATMAN and NICOLSON 1993). In addition, adhesion molecules expressed on metastatic cells may be responsible for the selective lodgment of the invading cells in specific organs. Indeed, YEATMAN and NICOLSON (1993) demonstrated, using an in vitro assay, that mouse B16 melanoma cells, selected by serial in vivo passages for brain or lung colonization, adhere at greater rates to the endothelium of their own target microvessels than do the parental cell lines or lines selected for other organ specificities. Lymphoma cells expressing the α4β7-integrin exhibited MAd-dependent adhesion to mucosal HEV (STRAUCH et al. 1994) and preferential homing to Peyer's patches and mesenteric lymph nodes (GOSSLAR et al. 1996).

A murine T cell lymphoma, designated LB, has been used in our laboratory as an experimental model for exploring the process of integrin- and CD44-dependent malignant dissemination in the context of the tumor's adhesiveness and homing properties. This article surveys our major in vivo and in vitro findings and attempts to evaluate them against corresponding observations related to normal physiological activities. For a more general perspective, we refer the reader to a few comprehensive review articles (DUSTIN and SPRINGER 1991; PICKER and BUTCHER 1992; LESLEY et al. 1993; SPRINGER 1994; NAOR et al. 1997).

Fig. 1A–C. The CD44 glycoprotein (**A**), its exon map (**B**) and examples of six alternatively spliced transcripts (**C**). **A** Protein structure. Using disulfide bonds, the NH_2-terminal of the molecule forms a globular domain, or three globular subdomains. The circle and the "downstream" ellipse represent areas that influence hyaluronate binding (PEACH et al. 1993; ZHENG et al. 1995). The *black track* inside the *circle* refers to a region displaying 30% homology with cartilage link protein and proteoglycan core protein, both showing HA binding ability. The *black track* at the NH_2-terminal (inside and outside the *circle*), transmembrane-spanning domain (23 amino acids) and cytoplasmic tail (70 amino acids) represents regions with 80%–90% interspecies homology. The alternatively spliced short cytoplasmic domain (3 amino acids) is nonproportionately represented by a *small bar*. The *lightly shaded track* in the *center* indicates the nonconserved membrane-proximal region, which display 35%–45% interspecies homology. The optional variable region, containing various combinations of variant exon products (see **C**) is inserted between amino acids 201 and 202 (mature protein) and marked by a zig-zag track. The full amino acid sequence of human and mouse CD44s is presented in ZHOU et al. 1989 and the nucleotide sequence (including the variable region) of human CD44 in SCREATON et al. 1992. *Filled circles*, potential N-linked glycosylation. *X*, areas rich in serine/threonine, possible sites for O-linked glycosylation (those of the variable region are arbitrarily assigned). *Filled diamonds*, potential sites for glycosaminoglycans (chondroitin sulfate, heparan sulfate) incorporation. *Open circles*, potential sites for phosphorylation (only part of the sites are depicted). The symbols on the standard part of the molecule mostly refer to mouse CD44 (ZHOU et al. 1989), whereas those of the variable region are based on information taken from both mouse and humans. **B** Exon map. The *filled circles* represent exons of the constant regions. *Open circles* represent variant exons that can be inserted by alternative splicing in the variable region. Note: exon V1 is not expressed in the human CD44. *LP*, leader peptide-encoding exon; *TM*, transmembrane-encoding exon. *CT*, cytoplasmic tail – encoding exons. **C** Examples of alternatively spliced transcripts. *1* and *2*, Standard CD44 with short and long cytoplasmic tails, respectively, which lack the entire variable region. *3*, pMeta-1 (CD44v4-7). Exons v4, v5, v6 and v7 are inserted in tandem between exons 5 and 17. *4*, pMeta-2 (CD44v6,7). Exons v6 and v7 are inserted between exons 5 and 17. pMeta-1 and pMeta-2 are known as "metastatic" CD44, because their cDNA confers, upon transfection, metastatic potential on nonmetastatic rat pancreatic adenocarcinoma cells (GÜNTHERT et al. 1991). Note that exon 16 is not expressed in pMeta-1 or pMeta-2. *5*, Epithelial CD44 (CD44v8-10), expressed predominantly on epithelial cells. Exons V8, V9 and V10 are inserted between exons 5 and 16. *6*, Keratinocyte CD44 (CD44V3-10), one of the largest CD44 molecules known. Exons V3 through V10 are inserted between exons 5 and 16. (Reproduced from NAOR et al. 1997)

2 LB Cells Use Different Migration Routes for Spleen and Lymph Node Invasion

LB is a T cell lymphoma (Lugasi et al. 1990) spontaneously originating in a BALB/c mouse (Ruggiero et al. 1985) that aggressively invades the spleen and lymph nodes following s.c. inoculation. The time required to kill 50% of the mice following i.p. injection of 100 and 1000 LB cells is 21 and 16 days, respectively, and 12 days or less for 10^4 and 10^5 cells (Zahalka et al. 1995). Flow cytometry reveals that LB cells express the following cell surface molecules: Thy-1, CD8, major histocompatibility complex (MHC) class I antigen (K^d and D^d), CD25 (interleukin-2 receptor), J11d, B2A2, CD44 and LFA-1 (CD11a/CD18; αLβ2). The lymphoma cells only slightly express CD4 and CD11b (Mac-1), but do not display CD3, CD5, class II antigens, gp70 retroviral protein, CD11c (p150, 95) and MEL-14 (Lugasi et al. 1990; Zahalka et al. 1993, 1995, and unpublished observations). LB cells express the β1- and β7-chain integrins, but do not display the α4- and α5-chain of these molecules. Therefore, VLA-4 and VLA-5 cannot be detected on the cell surface (Gosslar et al. 1996). Although CD3 is not expressed on the surface of LB cells, immunoprecipitation assays revealed the presence of CD3 γ-, δ- and ε-chains in the cell extract (Tzivion and Naor, unpublished observations), indicating the T cell origin of these cells. In addition, LB cells exhibit the insulin receptor, and their growth, both in vitro (Pillemer et al. 1992) and in vivo (Sharon et al. 1993), is highly dependent on physiological concentrations of insulin. Yet, the in vitro proliferation of the lymphoma cells is also stimulated by interleukin (IL)-2 and IL-4 as well as by growth hormone (unpublished observations).

Cell suspensions of spleen or lymph nodes taken from BALB/c mice subcutaneously inoculated with LB cells one day earlier, did not kill, following their transfer, naive recipient mice. This indicates that at that time the number of invading tumor cells in these organs was very small and, therefore, nonlethal. However, when spleen, but not lymph node, cells were transferred 4 days after lymphoma inoculation, they killed the recipient animals. In the lymph node, 7 days were required to generate a lethal dose of LB cells, yet the lethal effect of the spleen did not decline at that time, as demonstrated by the cell transfer experiment. This sensitive assay suggests that the lymphoma cells enter the spleen before they arrive the lymph nodes and that invasion of the spleen is not necessarily accomplished by tumor cells previously parked in the lymph node (Zahalka et al. 1993).

It should be emphasized, however, that 3 days after s.c. inoculation of LB cells into the animal's left flank, near the hind limb, histological examination revealed the presence of a few tumor cells in the subcapsular sinus of the axillary and brachial lymph nodes (Zahalka et al. 1995). This low number of LB cells is presumably not lethal in the above described transfer experiment. A different tactic (Gosslar and Holzmann, Technical University, Munich) involved the tagging of LB cells with β-galactosidase (β-gal), using retrovirus-mediated gene transfer, and the s.c. injection of one of the isolated labeled clones (G1) into the neck of syngenic BALB/c mice. Frozen, inguinal lymph node sections were prepared 10 days later and stained with the β-gal substrate 5-bromo-4-chloro-3-inodyl-β-D-galactopyranoside (x-gal). The $LB^{\beta\text{-gal}}$ cells stained blue. As seen in Fig. 2, the $LB^{\beta\text{-gal}}$ cells were concentrated mainly in the marginal zone of the lymph node, whereas only a few scattered cells were detected in the cortical or medullary areas. Similar results were obtained with other $LB^{\beta\text{-gal}}$ clones. Both the histological and the histochemical analyses show that s.c. inoculated LB cells first accumulate in the subcapsular sinus and then penetrate the deep regions of the lymph node, indicating that they invade the lymph nodes via the afferent lymphatics. Subcutaneously inoculated LB cells enter the spleen via the blood circulation. At 12 days after i.v. (rather than s.c.) injection, LB cells were detected in the spleen, but not in the lymph nodes (Zahalka et al. 1995). This is hardly surprising, in view of the fact that the lymphoma cells do not express the MEL-14 homing molecule (Zahalka et al. 1993), and, therefore, cannot bind to the lymph node HEV as normally do naive blood-borne lymphocytes, which display this surface molecule (Gallatin et al. 1983). Indeed, a binding assay of a frozen lymph node section (Butcher et al. 1979) revealed that LB cells do not adhere in vitro to the HEV (Zahalka et al. 1995), proving that they do not express other surface molecules implicated in HEV-associated entry into the lymph nodes. In this context, it should be recalled that mouse CD44 (expressed on LB cells), in contrast to human CD44, does not interact with HEV (Culty et al. 1990). It should be stressed, however, that at a later time, or under different experimental conditions, i.v. injected

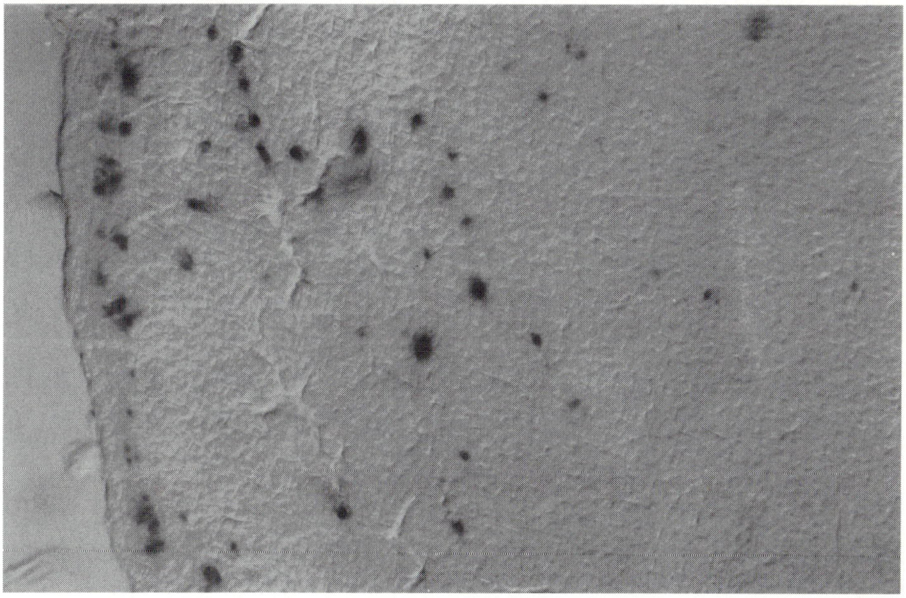

Fig. 2. Penetration of LB cells into the lymph node via the afferent lymphatics. LB cells were tagged with β-galactosidase using retrovirus-mediated gene transfer, cloned and subcutaneously injected into the neck of syngeneic BALB/c mice. Frozen, inguinal lymph node sections were prepared 10 days later and stained with x-gal. Histological analysis revealed that the tagged LB cells had accumulated mainly in the marginal zone of the lymph node, indicating that they invade this organ via the afferent lymphatics. ×400

LB cells may enter the lymph nodes through the afferent route after having disseminated in the lymphatics, or via the HEV due to up-regulation of cell surface HEV binding molecules. Sheep memory T lymphocytes, which, like LB cells, express high levels of CD44, but lack MEL-14, enter the lymph node through the afferent lymphatics (MACKAY et al. 1990), precisely like the lymphoma cells. It is tempting to speculate that in the mouse (and perhaps in sheep as well) CD44 is the key for entry into the lymph node via the afferent lymphatics, whereas CD18 is the key that opens the capillary gate of the spleen (see Sect. 3). In addition, the distinctive expression of MHC molecules on spleen – vs lymph node – infiltrating LB cells may also contribute to their discrete pattern of homing. However, this notion does not reconcile with the flow cytometry analysis (Fig. 3) of cells isolated from spleen and lymph nodes, which revealed that both cell types express the D^d antigen, but not the K^d antigen, whereas the parent LB cells express both class I products. Indeed, it has been shown (EISENBACH and FELDMAN 1991; FELDMAN and EISENBACH 1991) that the emergence of a metastatic phenotype is sometimes associated with a change in MHC expression.

Careful histological examination of blood sample smears did not reveal the presence of LB cells in the blood circulation (ZAHALKA et al. 1993), suggesting that

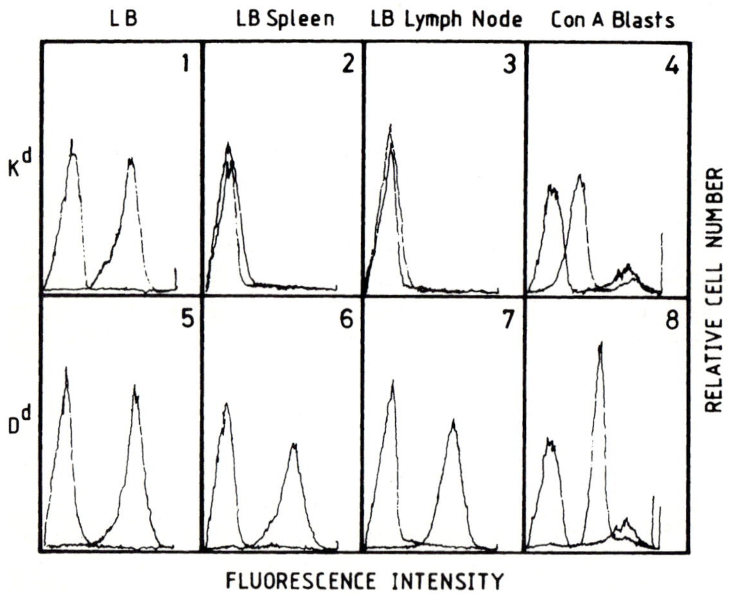

Fig. 3. Down-regulation of K^d class I antigen by spleen- and lymph node-infiltrating LB cells. Parental LB cells and LB cells isolated from spleen (LB spleen) and peripheral lymph node (LB lymph node) following their s.c. inoculation, were analyzed by flow cytometry, using anti-K^d and anti-D^d monoclonal antibodies (mAbs). Concanavalin A-induced blast cells served as a positive control. *Left* histogram in each panel, nonspecific binding of the indicator fluorescein-labeled antibody to the various cell types. *Right* histogram in each panel, specific binding of the mAb to the same cells. When two histograms match, specific binding is negligible or absent. Both K^d and D^d antigens were detected on parental LB cells, whereas only D^d antigen was present on LB cells isolated from the lymphoid organs

only a very small number of tumor cells leave the primary skin growth, later penetrating the lymphoid organs via the blood and afferent lymphatics and then massively proliferating there.

As mentioned earlier, LB cells do not express the α4-integrin chain (although they do express the β1- and β7-chains) and, consequently, do not display the intact α4β1- (VLA-4) and α4β7-integrins (GOSSLAR et al. 1996). As VLA-4 and α4β7 are implicated in the cell interaction with endothelium and the subsequent homing to lymphoid organs or inflamed tissues (PICKER and BUTCHER 1992), GOSSLAR and colleagues (1996) investigated how cell surface expression of α4-integrin influences the establishment of the lymphoma cells in various organs. To this end, LB cells were infected with the NTK-α4 recombinant retrovirus or with virus carrying an empty pNTK vector. The retrovirus-mediated gene transfer conferred uniform α4 expression on polyclonal or clonal α4 cell lines and, as a result, the cells displayed functional α4β1 (VLA-4) and α4β7 molecules, as indicated by flow cytometry and adhesion assays. After i.v. injection, there was a similar accumulation of ^{51}Cr-labeled LB-α4 and control LB-NTK cells in blood, intestine, skin, spleen, lung, liver and peripheral lymph nodes. LB-α4 cells, however, preferentially migrated to mesenteric lymph nodes and Peyer's patches when compared with LB-NTK cells, a phenomenon consonant with the finding that α4β7-integrin guides lymphoid cells to mucosal tissue (HU et al. 1992; HAMANN et al. 1994). Although LB-α4 cells migrate to lymphoid and nonlymphoid target organs at least as efficiently as their LB-NTK counterparts, establishment of the former in these organs (with the exception of bone marrow) is markedly reduced, as indicated by tumor cell proliferation assays or histological analysis. Impairment of the metastatic capacity of LB-α4 clones has also been observed (GOSSLAR et al. 1996). Notably, the ability of LB-α4 cells to form colonies in mesenteric lymph nodes and Peyer's patches was inhibited, despite the fact that these cells migrated more efficiently than did control LB-NTK cells to mucosal sites. This finding, together with the observation that both LB-α4 and LB-NTK cells exhibit the same in vitro proliferation rate and expand with the same efficiency in bone marrow, implies that the expression of α4 does not induce a general nonspecific proliferative defect in the lymphoma cells. If so, what is the mechanism that restrains the growth of LB-α4 cells in various target organs? Perhaps, signals delivered through α4-integrin molecules activate programmed cell death in LB cells, as demonstrated for chronically stimulated T cells coligated with anti-T cell receptor monoclonal antibody (mAb) and VCAM-1 (DAMLE et al. 1993). Alternatively, the α4-mediated signals may reduce the in vivo proliferation rate of the tumor. Consistent with the inhibitory effect of α4-integrins on metastasis formation is the observation that the expression of α4β1 is considerably lower in Burkitt lymphoma cells than in an Epstein-Barr virus-transformed B lymphoblastoid cell line (RINCON et al. 1992). An entirely different mechanism curbs the metastasis of murine B16 melanoma cells transfected with α4 cDNA and s.c. injected into C57BL/6 mice. Cell surface expression of α4β1-integrin prevents detachment of the melanoma cells from the primary growth and subsequent pulmonary metastasis, perhaps due to VLA-4-dependent homotypic adhesion (QIAN et al. 1994). In contrast, the metastasis of i.v. injected LB-α4 lymphoma cells in lymphoid and nonlymphoid organs is inhibited at a later stage, subsequent to tumor cell migration.

3 LB Cells Use Distinct Adhesion Molecules for Spleen and Lymph Node Invasion

As LB cells coexpress the LFA-1 integrin and the CD44 adhesive receptor (Zahalka et al. 1995), both implicated in cell migration and homing of normal lymphocytes, we conceived that they might be involved in lymphoid organ invasion by the tumor. This prediction was experimentally challenged by injecting anti-CD18 mAb, directed against the β2-chain of LFA-1 or anti-pan-CD44 mAb directed against the constant epitope shared by all CD44 isoforms into LB cell-inoculated mice (Zahalka et al. 1993, 1995). A quantity of 3×10^6 LB cells were s.c. inoculated into the left flank (close to the hind limb) of female BALB/c mice. Two hours later, the mice were administered a 50% ammonium sulfate fraction (500 μg protein) of anti-CD18 mAb or anti-CD44 mAb and the injections were repeated on alternate days until termination of the experiment (day 12). The anti-CD18 mAb was injected intravenously, whereas the anti-CD44 mAb was injected subcutaneously, near the remote front left axillary and brachial lymph nodes, as this route was found to be more efficient than i.v. injection. Isotype-matched mAbs (e.g. anti-CD4 mAb) or mAbs directed against nonrelevant LB cell surface molecules served as control. We found that i.v. injection of anti-CD18 mAb (M18/2) or equivalent amounts of its $F(ab')_2$ fragments reduced the number of lymphoma cells invading the spleen by at least two orders of magnitude: from hundreds of thousands to several hundreds. The anti-CD18 mAb did not, however, prevent peripheral lymph node invasion (Zahalka et al. 1993). By contrast, anti-CD44 mAb (IM7.8.1) or equivalent amounts of its $F(ab')_2$ or Fab' fragments reduced the number of lymphoma cells infiltrating the peripheral lymph nodes by two orders of magnitude after s.c. injection of these reagents, whereas there was no effect on spleen invasion (Zahalka et al. 1995). The fact that $F(ab')_2$ antibody fragments also inhibited lymphoid organ invasion by LB cells indicates that the antibody did not eliminate the lymphoma by complement-dependent lysis or by antibody-dependent cellular clearance. Furthermore, the finding that the Fab fragments inhibited invasion excludes the possibility that the effect is mediated by modulation of cross-linked cell surface molecules. It should be noted, however, that anti-CD18 and anti-CD44 mAbs effectively blocked invasion of spleen and lymph node by LB cells only when injected 1 and 2 days, respectively, after tumor inoculation (Fig. 4). When both antibodies were simultaneously injected, lymph node invasion was inhibited, whereas spleen invasion was restored (Zahalka and Naor 1994). A possible explanation for this surprising phenomenon, will be presented in Sect. 4. The finding that α4-integrin expression strongly reduced spleen and lymph node metastases by LB-α4 cells (see previous section), may mean that the suppressive effect dominates

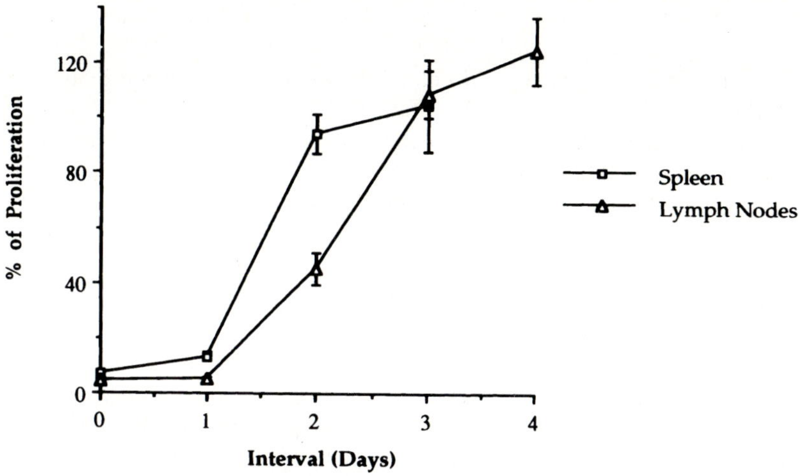

Fig. 4. Inhibition of LB cell infiltration into spleen and lymph nodes by anti-CD18 and anti-CD44 monoclonal antibodies (mAbs) injected at different intervals after LB cell inoculation. A quantity of 3×10^6 LB cells were s.c. injected into BALB/c mice (3 per group). After 2 h (time 0 on the graph), anti-CD18 mAb was injected i.v. or anti-CD44 mAb was injected s.c. into different groups of mice. Both antibodies were precipitated with 50% ammonium sulfate and 0.5 mg protein was injected into each mouse. The same amount of protein was injected every other day until termination of the experiment (day 12). Other groups of mice were subjected to the same experimental protocol, but antibody injection was begun 1, 2, 3 or 4 days after tumor inoculation. Spleen invasion in anti-CD18 mAb injected mice and lymph node invasion in anti-CD44 mAb injected mice were determined by measuring LB cell proliferation ([^{3}H]thymidine uptake) in cell suspensions obtained from the spleen and lymph nodes of these animals. The results are expressed as percent proliferation of spleen (116000 cpm) and lymph node (68000 cpm) cells of mice inoculated s.c. with lymphoma cells alone. The anti-CD18 mAb (*squares*) and anti-CD44 mAb (*triangles*) inhibited invasion of the spleen (*squares*) and peripheral lymph node (*triangles*) by proliferating LB cells only when injected 1 and 2 days, respectively, after tumor inoculation

the CD18- and CD44-mediated adhesive activities promoting lymphoma invasion of these organs. The inhibitory effect of the antibodies was established according to: (1) histopathological findings in the lymphoid organs, (2) the reduced uptake of [^{3}H]thymidine by LB cells populating the lymph node and spleen cell suspensions obtained from the lymphoma-inoculated and antibody injected mice, indicating decreased proliferation of tumor cells in these suspensions (evidence that the proliferating cells are LB cells and not local cells of the invaded organs is presented in Zahalka et al. 1993), and (3) the low number of tumor cells in suspensions of lymph node and spleen transferred from LB cell-inoculated and antibody-injected mice into naive recipients, based on recipient survival. In conclusion, our findings show that the CD18 and CD44 epitopes are used by LB cells as key codes for differential homing of the tumor to the spleen or lymph nodes. As LB cells isolated from both organs coexpress CD18 and CD44, as do the parental cells obtained from the local growth (Vogt Sionov and Naor 1997), it is conceivable that the predominant expression of the CD18 ligand in the spleen, and of the CD44 ligand in the lymph node, dictates the differential adhesion-dependency of lymphoid organ invasion by the tumor (analogous to a lock that dictates which key in a set of available keys is used). We previously emphasized that normal leukocytes also use distinct adhesion molecules for lodgment in different lymphoid organs, thereby underlying the similarity between lymphoma dissemination and normal migration of hematogenous cells.

The influence of integrins or CD44 on the neoplastic process has been demonstrated in many experimental models by cDNA transfection experiments, evaluation of the ability of relevant mAbs or soluble proteins to affect tumor growth and, in clinical studies, by the correlation between adhesion molecule expression and the tumor progression. Transfection of human α5- and β1-integrin cDNAs suppressed the tumorigenicity of Chinese hamster ovary cells (Giancotti and Ruoslahti 1990), and expression of α4β1-integrin in B16 melanoma reduced matrigel invasion in vitro and suppressed pulmonary metastasis in vivo (Qian et al. 1994). By contrast, expression of α2β1-integrin in human rhabdomyosarcoma cells enhanced experimental and spontaneous metastases in nude mice (Chan et al. 1991). In concurrence with our observations, it has been reported that mAbs directed against LFA-1 subunits (α [CD11a] or β [CD18] chains) or against its counterpart ICAM-1 (CD54) inhibit the dissemination of lymphoma or myeloma cells in animal models (Harning et al. 1993; Huang et al. 1995; Rocha et al. 1996). Reduced expression of α4β1-integrin was found in Burkitt lymphoma (Rincon et al. 1992), whereas up-regulation of β3 integrins was detected in metastatic melanoma (Albelda et al. 1990). It has been shown, however, that many high grade human lymphomas do not express LFA-1 molecules (Clayberger et al. 1987), suggesting that the selective pressure favoring LFA-1 expression as a tool for tumor expansion may be outweighed by the selective pressure against LFA-1 expression, preventing immunological destruction of the malignant cells.

We have already mentioned that transfection with the CD44-containing V6 exon cDNAs (pMeta-1) conferred metastatic behavior on a nonmetastatic pancreatic adenocarcinoma cell line (Günthert et al. 1991; Rudy et al. 1993). Melanoma (Bartolazzi et al. 1994) or Burkitt lymphoma (Sy et al. 1991; Bartolazzi et al. 1995; Walter et al. 1995) human cell lines transfected with CD44s cDNA exhibited accelerated tumor growth in immunodeficient mice. Injection of anti-CD44v mAb retarded the killing of rats by metastatic pancreatic carcinoma (Seiter et al. 1993). The local growth and metastatic spread (especially into the lung) of a human melanoma cell line s.c. inoculated into immunodeficient mice was inhibited by injections of mAb directed against the CD44 constant region (Guo et al. 1994). Continuous infusion of soluble CD44 (CD44-immunoglobulin [Ig] fusion protein) through an osmotic pump prevented tumor development following s.c. injection of B16 melanoma cells into mice (Bartolazzi et al. 1994).

A correlation between CD44v expression and the stage of tumor progression has been described in some human malignant diseases (e.g., colorectal cancer; Wielenga et al. 1993). An inverse correlation between CD44 expression and poor prognosis has also been reported (e.g., in neuroblastoma) as has the absence of either correlation (reviewed in Naor et al. 1997). However, in many cases, conflicting observations make it impossible to draw any conclusions about CD44 association with neoplastic diseases (reviewed in Naor et al. 1997). In non-Hodgkin's lymphomas (NHL), immunohistochemical studies showed that the majority, or at least a significant part, of tumor specimens from patients classified as intermediate/high grade express CD44 variants, mostly those containing exons V3, V6 or V9 (KOOPMAN et al. 1993; TERPE et al. 1994; STAUDER et al. 1995). Reverse transcriptase-polymerase chain reaction (RT-PCR) analysis revealed a more complicated pattern (a greater number and larger size) of CD44 variant transcripts in tumor specimens of high grade NHL patients than in low grade ones (STAUDER et al. 1995). Irrespective of grade type, the survival of NHL patients with CD44V6-positive tumors is shorter than that of those with CD44V6-negative tumors (RISTAMÄKI et al. 1995; STAUDER et al. 1995).

4 Different Strategies Exercised by LB Cells for Their Establishment in Spleen and Lymph Nodes

LB cells form heterotypic aggregates with splenocytes, but not with lymph node cells. Maximal in vitro aggregate formation was detected when spleen and lymphoma cells were coincubated at an 8:1 ratio (ZAHALKA et al. 1993; ZAHALKA and NAOR 1994). Spontaneous aggregate formation was observed when spleen (but not lymph node) cell suspensions of LB cell-inoculated mice were cultivated in vitro for 16 h. The ex vivo cell aggregation was maximal 7–8 days after tumor inoculation. Using mAbs that preferentially bind to splenocytes or LB cells, we showed that both the in vitro and the ex vivo disrupted aggregates consisted of a mixture of splenic and tumor cells (ZAHALKA et al. 1993; ZAHALKA and NAOR 1994). The lymphoma cells tended to form aggregates with splenic T cells, rather than with B cells or other non-T cells, as demonstrated by incubation of LB cells with subpopulations of spleen cells or by flow cytometric analysis of disrupted aggregates. The involvement of T cells in aggregate formation was also proved by the weak ability of splenocytes from thymusless nude BALB/c mice to form aggregates with LB cells. The assumption that the T cell-LB cell interaction is essential for tumor growth in the spleen is supported by the observation that proliferating lymphoma cells were hardly detected in the spleens of tumor-inoculated nude mice, whereas such cells were present in their lymph nodes (ZAHALKA and NAOR 1994).

The formation of aggregates between splenocytes and LB cells was prevented when anti-CD18 mAb, but not control mAb, was included in the cell mixture. Similarly, injection of anti-CD18 mAb (but not control mAb) into LB cell-inoculated mice markedly reduced the formation of spontaneous ex vivo aggregates in cell suspension of their spleens. These findings show that aggregation between splenic T cells and LB cells is CD18-dependent (ZAHALKA et al. 1993; ZAHALKA and NAOR 1994).

As LB cells express the LFA-1 integrin, comprising αL-(CD11a) and β2-(CD18) chains, we also tested the ability of anti-CD11a mAb (M7/14) to prevent spleen invasion by LB cells (ZAHALKA et al. 1993), as well as in vitro (ZAHALKA et al.

1993) and ex vivo (ZAHALKA and NAOR 1994) formation of aggregates between splenocytes and LB cells. In contrast to anti-CD18 mAb, anti-CD11a was ineffective in both aspects. Anti-ICAM-1 (YN1/1.7) mAb also failed to inhibit spleen invasion by LB cells and aggregate formation between splenocytes and lymphoma cells (ZAHALKA et al. 1993). This is not surprising, as the LFA-1 α-subunit binds to ICAM-1 (JOHNSTON et al. 1990). We assessed the bioactivity of anti-CD11a mAb by testing its ability to block the function of cytotoxic T lymphocytes (CTLs), generated in mixed lymphocyte culture. We found that when anti-CD11a mAb was added to a mixture of cytotoxic cells and target cells it inhibited CTL activity, whereas anti-CD18 mAb was much less effective (ZAHALKA et al. 1993). This finding suggests that an epitope located on the β-chain of LFA-1 is involved in LB cell dissemination, whereas an epitope situated on the α-chain is implicated in CTL recognition of the target cell, as reported earlier (SANCHEZ-MADRID et al. 1983). The CD18 β-chain molecule, which affects tumor spread and aggregate formation, may associate with different α-chains. As αM (CD11b) is marginally expressed and αX (CD11c) is not expressed on LB cells (ZAHALKA et al. 1993), they most likely are other members of this family or, less probably, unrelated molecules. Immunoprecipitation of LB cell extracts with anti-CD18 mAb and subsequent gel electrophoresis revealed, in addition to the αL-(180 kDa) and β2-(95 kDa) chains, extra bands at 150 kDa and at 32–35 kDa. These disappeared after washing the immunoprecipitate with LiCl solution, indicating that they were loosely bound to the complex and, therefore, did not represent classical α-chains. Yet, these extra molecular species could be important for LB cell dissemination and/or heterotypic aggregation. The possibility that certain epitope(s) located on the αL-chain of LB cell LFA-1 are essential for tumor dissemination and/or aggregation, but that these epitopes are not recognized by M7/14 anti-CD11a mAb, also must be taken into account.

Coinjection of anti-CD18 (i.v.) and anti-CD44 (s.c.) mAbs into BALB/c mice s.c. inoculated with LB cells antagonized the spleen invasion blocking effect of anti-CD18 mAb and reduced its ability to inhibit splenic ex vivo aggregate formation. Whereas the injection of both antibodies partially restored spleen invasion by the lymphoma, the treatment did not influence the inhibition of lymph node infiltration by the anti-CD44 mAb (ZAHALKA and NAOR 1994). These results are best interpreted by assuming that interaction of anti-CD44 mAb with cell surface CD44 augments the expression of the CD18 molecules on the LB cells, or alters their configuration, rendering them less susceptible to the inhibitory effect of anti-CD18 mAb. According to flow cytometry analysis, however, expression of the lymphoma cell CD18 molecules was not enhanced after treatment with anti-CD44 mAb. Therefore, it is most likely that interaction of anti-CD44 mAb with CD44 changes the configuration of the CD18 molecules, thus reducing the ability of anti-CD18 mAb to inhibit LB cells from binding to splenic T cells, an event probably mandatory for lodgment and proliferation of the tumor in this organ. In agreement with our findings, it has been shown in other systems (KOOPMAN et al. 1990; RODRIGUES et al. 1992; FUNARO et al. 1994) that anti-CD44 mAbs activate homotypic T cell aggregation, which is LFA-1/ICAM-dependent, deduced from the finding that antibodies against the integrin disturb aggregate formation.

Table 1. Differences and similarities between spleen- and lymph node-infiltrating LB cells

Parameter	Spleen-infiltrating LB cells	Lymph node-infiltrating LB cells
1. Time required to reach the lymphoid organ	4 days	7 days
2. Route of entry into the lymphoid organ	Blood circulation	Afferent lymphatics
3. Ability to form in vitro and ex vivo aggregates with lymphoid organ cells	Yes	No
4. Predominant binding target	T cells	Extracellular matrix component?
5. Adhesion molecules expressed	CD18, CD44	CD18, CD44
6. Class I antigens expressed	D^d	D^d
7. Predominant adhesion molecule required for organ invasion	CD18 (β2 chain integrin)	CD44
8. Adhesion receptor ligand	Unknown	Unknown

The preferential aggregation of LB cells with splenic T cells and the lack of such aggregate formation in the lymph nodes lead us to surmise that the mechanisms responsible for LB cell lodgment in spleen and lymph node are distinct and obey different signals. We suggest that invasion of the spleen is associated with the CD18-dependent interaction between the tumor cells and the splenic T cells. This may confer some growth advantage on the former, as T cells supply the IL-2 and IL-4 necessary for the lymphoma's proliferation (Lugasi et al. 1990 and unpublished data). Indeed, in nude mice, the T cell-deficient splenocytes showed a reduced capacity for forming aggregates with LB cells, and they barely supported proliferation of the tumor (Zahalka and Naor 1994). By contrast, the CD44-dependent lodgment of LB cells in the lymph node is presumably associated with binding of the tumor to ECM components, which may also deliver the mitogenic signal to the invading cells. Table 1 summarizes the differences and similarities between spleen- and lymph node-infiltrating LB cells.

5 The LB Cell CD44 Ligand

The ability of CD44 to bind various ligands (e.g., hyaluronic acid, fibronectin, collagen and osteopontin) has been attributed to the multistructural nature of this molecule, as mentioned earlier in the Introduction. Identification of the CD44 ligand is not only academically important, but also has practical implications, because interference with the receptor-ligand interaction may influence the receptor's pathological and physiological activities. Determination of the LB cell CD44 ligand and an understanding of its mode of interaction with the receptor have been the thrust of our subsequent research efforts.

As hyaluronic acid (HA) is the principal ligand of CD44 (reviewed in Lesley et al. 1993 and Naor et al. 1997), the interaction of this molecule with LB cells has been intensively investigated. HA is a ubiquitous polysaccharide (glycosaminoglycan) consisting of a linear polymer of repeating disaccharide units with the structure (D-glucuronic acid [1-β-3] N-acetyl-D-glucosamine $[1\text{-}\beta\text{-}4])_n$ (LAURENT and FRASER 1992). This large molecule (molecular mass, 10^6–10^7 daltons) is an important component of the ECM, fills the intracellular spaces, provides cellular support and a water-filled compartment, and regulates cell-cell adhesion as well as the cell's spatial orientation and traffic (HARDINGHAM and FOSANG 1992; LAURENT and FRASER 1992). Therefore, it is not surprising that HA enhances tumor invasiveness and metastasis (TOOLE et al. 1979; ZHANG et al. 1995). [^{3}H]thymidine-labeled LB cells do not bind to HA immobilized to plastic, unless they are activated by phorbol 12-myristate 13-acetate (PMA). Other glycosaminoglycans (heparin, heparan sulfate and chondroitin sulfate) do not bind activated LB cells. Anti-CD44 mAb and its $F(ab')_2$ or Fab fragments, but not the same amount of anti-CD18 or isotype matched anti-CD4 mAb, prevent the binding of PMA-activated LB cells to immobilized HA, indicating that the binding of the lymphoma to this ligand is CD44-dependent. Immobilized HA pretreated with hyaluronidase, but not with heparinase or chondroitinase AC, failed to bind activated LB cells, again stressing the specificity of the adherence (ZAHALKA et al. 1995).

LB cells do not bind soluble, fluorescein-labeled hyaluranic acid (Fl-HA), even after stimulation with low concentrations of PMA (VOGT SIONOV and NAOR 1997). Negatively charged carbohydrate groups of the CD44 glycoprotein may interfere with the interaction between its positively charged amino acids (arginine and lysine) and the negatively charged groups of HA. Hence, removal of the negatively charged sialic acid from LB cell CD44 by neuraminidase, or treatment with tunicamycin, to prevent N-glycosylation of the cell receptor, may allow HA binding. Although such treatment enabled other cell lines, whose CD44 receptor was initially inactive, to bind soluble HA (KATOH et al. 1995; LESLEY et al. 1995), deglycosylated LB cells remained incapable of binding the ligand. However, we found that LB cells display Fl-HA binding after stimulation with low concentrations of PMA and treatment with tunicamycin, whereas incubation with either one of these reagents was ineffective (deglycosylation was proven by gel electrophoresis of CD44 immunoprecipitated from LB cell extract). A similar finding of HA binding was observed when LB cells were treated with neuraminidase following activation with the phorbol ester. Again, tumor cells treated with neuraminidase alone or PMA alone did not bind Fl-HA. Another set of experiments demonstrated that LB cells cultivated in glucose-deprived medium and stimulated with low concentrations of PMA bound soluble HA, whereas corresponding cells cultured in glucose-containing medium did not show this property (Rochman, Naor and Ish-Shalom, unpublished data). In conclusion, our results suggest that two events are required for HA binding by the lymphoma cells: an increase in the net positive charge of the cell CD44 receptor, achieved by deglycosylation, and protein kinase C (PKC) activation, induced by phorbol ester. Activation of PKC may influence the interaction between the CD44 receptor and the cytoskeleton, which has been shown to be essential for HA binding to some (BOURGUIGNON et al. 1993; LIAO et al. 1993; LOKESHWAR et al. 1994; GALLUZZO et al. 1995), but not all (MURAKAMI et al. 1994; PERSCHL et al. 1995; UFF et al. 1995) cell types. Since deglycosylation affects not only the CD44 of LB cells, but also other cell surface glycoproteins, including those that may cooperate with CD44 in HA binding, it is impossible, under such circumstances, to specify the target molecule(s) of the deglycosylation procedure. It has been shown, however, by other investigators (KATOH et al. 1995) that removal of sialic acid by neuraminidase from CD44-Ig fusion protein absorbed to protein A-conjugated Sepharose beads enhances HA binding to the coated beads, indicating the direct interference of the CD44 carbohydrate moiety with HA binding.

If lymph node lodgment by LB cells is dependent on the interaction between LB cell CD44 and the hyaluronan of the organ ECM, we should expect LB cells isolated from the lymph nodes to have acquired the HA-binding capacity. The in vivo acquisition of this property by LB cell CD44 may be dependent on the replacement of factors detected in our in vitro studies (PMA and deglycosylation) by other factors (e.g., cytokines), as well as on phenotypical changes in cell surface CD44. To test this hypothesis, we compared the phenotype and the HA-binding capacity of LB cells isolated from culture or obtained directly from local growths, remote axillary lymph nodes and spleen 12 days after s.c. inoculation of the tumor into BALB/c mice. Flow cytometry analysis revealed that the LB cells derived from all these sites display the same intensity of pan-CD44 (detected by anti-CD44 mAb directed against the CD44 constant region shared by all CD44 isoforms). Using anti-V4 and V6-specific mAbs (kindly provided by Dr. J. Moll, Karlsruhe Research Center) we found that the lymphoma cells from all four sites slightly, but clearly, express V6-containing CD44 variants and hardly express V4-containing isoforms. The expression of all three isoforms was enhanced after PMA activation. Similarly to LB cells isolated from culture, those obtained from a local tumor, a peripheral lymph node and spleen did not adhere to immobilized HA unless activated by PMA. Neither did the lymphoma cells isolated from all four sources bind soluble Fl-HA, even after activation with the phorbol ester and/or treatment with hyaluronidase, as indicated by flow cytometry. The failure to interact with Fl-HA after enzyme treatment proves that the inability to bind the ligand is not due to possible masking of the CD44 receptor by prebound, tissue-derived HA. We will see later that hyaluronidase treatment can remove excess of soluble HA prebound to cell surface CD44 without reducing the receptor capacity to subsequently bind Fl-HA. In conclusion, the tested phenotype of the LB cells was not changed after their lodgment in the lymph node. Even more important, the tumor cells isolated from the lymph node remained incapable of binding hyaluronan, suggesting that interaction with a different ligand is required for docking in the organ parenchyma (VOGT SIONOV and NAOR 1997).

To further substantiate our experimental approach, an HA-binder LB cell line (designated HA9) was generated from an HA-nonbinder LB cell clone (LB 2.3) by nine cycles of selection for cells adhering to immobilized HA. HA9 cells express pan-CD44 and CD44 isoforms containing V4 or V6 exon products more intensively than do the parental LB cells. The expression of all three isoforms was further enhanced by stimulation with PMA. HA9 cells adhered to immobilized HA (but not to chondroitin sulfate) and bound Fl-HA from the solution even in the absence of PMA activation. The binding of soluble Fl-HA to HA9 cells was partially blocked by anti-pan-CD44 mAb, but not by anti-V4 or anti-V6 mAbs. Excess soluble HA did not prevent pan-CD44 mAb from binding to the tumor cells, suggesting that the pan-CD44 epitope and the HA binding site are located at different positions on the HA9 cell CD44 receptor and that anti-CD44 mAb allosterically inhibits HA from binding to the cell CD44. Adherence of HA9 cells to immobilized HA was inhibited by excess soluble HA or by anti-CD44 mAb, but not by anti-CD18 mAb, indicating that binding of the ligand is CD44-dependent. HA9 cells incubated with excess soluble HA to mask their CD44 receptor and then treated with hyaluronidase to remove the prebound ligand efficiently bound Fl-HA, as indicated by flow cytometry. HA9 cells that were first incubated with Fl-HA and then treated with hyaluronidase did not display the ligand binding, whereas cells that were first treated with the enzyme and then incubated with Fl-HA exhibited normal HA binding. These findings prove that hyaluronidase can remove prebound HA and directly bound Fl-HA from the CD44 receptor and that the digestion procedure does not hamper the receptor binding capacity.

HA9 cells isolated from culture or obtained from local growths, remote axillary lymph nodes and spleen displayed similar levels of pan-CD44 and of CD44 variants containing V4 and V6 exon products. In contrast to the corresponding LB cells, HA9 cells isolated from all four sites similarly adhered to immobilized HA and bound soluble Fl-HA from the solution even in the absence of PMA activation (Vogt Sionov and Naor 1997). These findings demonstrate that, like the parental LB cells, HA9 cells change neither their tested CD44 profile nor their HA binding capacity during the metastatic cascade (Vogt Sionov and Naor 1997). The results suggest, but do not prove, that both cell types are already precommitted to lymphoid organ invasion at the initial malignant phase of local growth. The alternative possibility that both cell infiltrates acquire alterations in phenotype, nondetectable in our assay protocol, should also be taken into account.

It should be further emphasized that the LB cells were analyzed after lodging in the lymph nodes, and not during their migration in the blood and lymph circulation. Therefore, we cannot rule out the possibility that, upon migration, LB cells temporarily acquire the HA-binding capacity, which is lost after they infiltrate the lymph node, as previously suggested by our group (Zahalka et al. 1995). In the earlier experiments we demonstrated the partial retardation of LB cell infiltration into lymph nodes when 150 U hyaluronidase were injected adjacent to the lymph node, but not when lower (75 U) or higher (300 U) doses were administrated. The inconsistency of these findings (Vogt Sionov and Naor 1997) has led us to assume that the dose-dependent effect of hyaluronidase previously observed may be due to a delicate balance between the ability of hyaluronidase to enhance lymph node metastasis (by destroying matrix resistance) and the enzyme's ability to inhibit metastasis (possibly by targeting and tuning the enzymatic activity to the tumor migration phase). If the CD44-HA interaction is crucial for lymphoid organ invasion, we should expect more rapid and efficient dissemination of HA9 cells able to bind HA than of the LB parental cells, which do not bind this ligand. However, the opposite results were obtained: 10–12 days were required for LB cells to populate the spleen and lymph nodes of syngeneic BALB/c mice vs the 18–22 days necessary for HA9 cells, whose proliferation rate in the organ was also lower than that of the parental cells (VOGT SIONOV and NAOR 1997). Despite their markedly different in vivo growth rates, the two cell populations displayed almost identical division rates in vitro, proving that they do not differ in their intrinsic proliferating potential. Hence, the tight interaction between HA9 cells and hyaluronan may slow the release of cells from the primary growth or retard their migration in the blood and lymph circulation. In addition, or alternatively, the immunological resistance of HA9 cells might be stronger than that of parental LB cells, causing a delay in the homing of the former to the lymphoid organs. In either case, this finding underlines that the CD44-HA interaction is not essential for LB lodgment in the lymphoid organs.

Other groups have also demonstrated that not all CD44-mediated cell activities are HA-dependent, including cell homing to lymph node (CULTY et al. 1990) and thymus (WU et al. 1993; PATEL et al. 1995), as well as binding of erythroid leukemic cells to hematopoietic supportive cells (SUGIMOTO et al. 1994). In this context, SLEEMAN and his colleagues (1996) reported that a rat tumor cell line transfected with hyaluronidase cDNA, whose product prevented the cells from interacting with hyaluronan, was as invasive as the wild-type parental cells, indicating that CD44 interaction with HA is not always essential for tumor metastasis. Furthermore, in some cases different adhesion molecules may exhibit redundant functions, so that the loss of one type of molecule can be functionally replaced by another. Indeed, DRIESSENS and colleagues (1995) showed that, although the HA-binding capacity of mouse lymphosarcoma cells is abolished by knocking out (by homologous recombination) the CD44 gene, this does not interfere with their local growth and metastatic spread, which are possibly mediated by integrins also expressed on the same tumor cells.

6 Conclusions

Malignant activity is a deflection of normal physiological functions, generated by genetic alterations and the consequent deterioration of check and balance mechanisms. Despite this distortion, many elements of similarity between normal and tumor cells are preserved. One of these is the adhesion molecule-dependent process of cell homing to lymphoid organs, as demonstrated in our studies on LB T cell lymphoma. This similarity constitutes an obstacle to clinical strategies of therapy, which are highly dependent on the ability to discriminate between normal and malignant processes. In this respect, the applicability of our experimental approach to clinical settings is limited, because blocking of lymphoma dissemination with anti-CD18 and anti-CD44 mAbs may interfere with the CD18- and CD44-dependent defense mechanisms of normal leukocytes. However, if tumor cells express modified versions of the adhesion molecules displayed by normal cells, a window for therapeutic intervention may be available, as the modified molecules could be used as specific targets for therapy. The possibility to detect distinct versions of integrins or CD44 is more promising in neoplastic cells than in their normal counterparts, since proliferating tumor cells are more susceptible to genetic alterations. If any of these genetic changes afford the tumor a biological advantage, a process of natural selection and stabilization would be expected in those cells expressing the modified molecules. From the practical aspect, future research should, therefore, focus on the detection of tumor-specific integrins or tumor-specific CD44. In this context, the CD44 receptor provides more opportunities, as differential alternative splicing generates many versions of this molecule. To date, 20 different CD44 isoforms are known but, theoretically, hundreds of isoforms can be formed by differential utilization of variant exons (van Weering et al. 1993). Based on the arguments presented above, the alternative splicing machinery should generate not only a richer variant repertoire in proliferating tumor cells than in normal cells, but also tumor-specific genetic alterations in the variants themselves (e.g., "illegitimate" insertion of introns, Matsumura et al. 1995). If some of these CD44 variants are exclusively expressed on tumor cells, they could be used as specific targets for therapy or diagnosis. Initially, CD44-containing V6 exon products were considered "metastatic" CD44 because some advanced human tumors (including lymphomas) preferentially express this entity (reviewed in Naor et al. 1997). Even more striking, anti-V6 mAb retarded the progression of rat pancreatic adenocarcinoma (Günthert et al. 1991). However, it was later found that immunologically activated lymphocytes express V6-encoded epitopes as well (Arch et al. 1992), making the V6 exon product a less attractive target. This disappointing finding should not discourage, but rather stimulate, the search for tumor-specific CD44 variants, as well as for other tumor-specific adhesion molecules (e.g., new α-chains of $\beta 2$-integrin or new members of the ICAM family). Identification of novel adhesion molecule ligands can offer another set of targets for therapy, provided that the cancer cells preferentially interact with them. The polymorphic structure of CD44 potentiates interactions with many ligands, some of them already known, others yet to be identified (e.g., the natural ligand of LB cell CD44). If some of these ligands interact with cancer cell CD44 isoforms, they could be targeted by antagonizing reagents, such as antibodies or competing analogues that may prevent the interaction. Paradoxically, the selection of CD44 variants that assure the tumor of survival advantage might also prove disadvantageous, since they could serve as specific targets for therapy.

Acknowledgements. The research described here was supported by the Society of Research Associates of the Lautenberg Center, Deutsche Krebshilfe, Dr. Mildred Scheel Stiftung and The Focused Giving Program, Johnson & Johnson. We thank Dr. Alexandra Mahler for her editorial assistance.

References

Albelda SM, Mette SA, Elder DE, Stewart RM, Damjanovich L, Herlyn M, Buck CA (1990) Integrin distribution in malignant melanoma: association of the β_3 subunit with tumor progression. Cancer Res 50:6757–6764

Arch R, Wirth K, Hofmann M, Ponta H, Matzku S, Herrlich P, Zöller M (1992) Participation in normal immune responses of a metastasis-inducing splice variant of CD44. Science 257:682–685

Bartolazzi A, Peach R, Aruffo A, Stamenkovic I (1994) Interaction between CD44 and hyaluronate is directly implicated in the regulation of tumor development. J Exp Med 180:53–66

Bartolazzi A, Jackson D, Bennett K, Aruffo A, Dickinson R, Shields J, Whittle N, Stamenkovic I (1995) Regulation of growth and dissemination of a human lymphoma by CD44 splice variants. J Cell Sci 108:1723–1733

Bourguignon LYW, Lokeshwar VB, Chen X, Kerrick WGL (1993) Hyaluronic acid-induced lymphocyte signal transduction and HA receptor (GP85/CD44)-cytoskeleton interaction. J Immunol 151:6634–6644

Butcher EC (1991) Leukocyte-endothelial cell recognition: three (or more) steps to specificity and diversity. Cell 67:1033–1036

Butcher EC, Scollay RG, Weissman IL (1979) Lymphocyte adherence to high endothelial venules: characterization of a modified in vitro assay, and examination of the binding of syngeneic and allogeneic lymphocyte populations. J Immunol 123:1996–2003

Chan BMC, Matsuura N, Takada Y, Zetter BR, Hemler ME (1991) In vitro and in vivo consequences of VLA-2 expression on rhabdomyosarcoma cells. Science 251:1600–1602

Clayberger C, Medeiros LJ, Link MP, Warnke RA, Wright A, Koller TD, Smith SD, Krensky AM (1987) Absence of cell surface LFA-1 as a mechanism of escape from immunosurveillance. Lancet 2:533–536

Culty M, Miyake K, Kincade PW, Silorski E, Butcher EC, Underhill C (1990) The hyaluronate receptor is a member of the CD44 (H-CAM) family of cell surface glycoproteins. J Cell Biol 111:2765–2774

Damle NK, Klussman K, Leytze G, Aruffo A, Linsley PS, Ledbetter JA (1993) Costimulation with integrin ligands intercellular adhesion molecule-1 or vascular cell adhesion molecule-1 augments activation-induced death of antigen-specific $CD4^+$ T lymphocytes. J Immunol 151:2368–2379

Driessens MHE, Stroeken PJM, Rodriguez Erena NF, van der Valk MA, van Rijthoven EAM, Roos E (1995) Targeted disruption of CD44 in MDAY-D2 lymphosarcoma cells has no effect on subcutaneous growth or metastatic capacity. J Cell Biol 131:1849–1855

Dustin ML, Springer TA (1991) Role of lymphocyte adhesion receptors in transient interactions and cell locomotion. Annu Rev Immunol 9:27–66

Eisenbach L, Feldman M (1991) Tumor cell surface determinants in host immunity against metastases. Cancer Biol 2:179–188

Feldman M, Eisenbach L (1991) MHC class I genes controlling the metastatic phenotype of tumor cells. Cancer Biol 2:337–346

Funaro A, Spagnoli GC, Momo M, Knapp W, Malavasi F (1994) Stimulation of T cells via CD44 requires leukocyte-function-associated antigen interactions and interleukin-2 production. Hum Immunol 40:267–278

Gallatin WM, Weissman IL, Butcher EC (1983) A cell-surface molecule involved in organ-specific homing of lymphocytes. Nature 304:30–34

Galluzzo E, Albi N, Fiorucci S, Merigiola C, Ruggeri L, Tosti A, Grossi CE, Velardi A (1995) Involvement of CD44 variant isoforms in hyaluronate adhesion by human activated T cells. Eur J Immunol 25:2932–2939

Giancotti FG, Ruoslahti E (1990) Elevated levels of the $\alpha_5\beta_1$ fibronectin receptor suppress the transformed phenotype of chinese hamster ovary cells. Cell 60:849–859

Gosslar U, Jonas P, Luz A, Lifka A, Naor D, Hamann A, Holzmann B (1996) Predominant role of α4-integrins for distinct steps of lymphoma metastasis. Proc Natl Acad Sci USA 93:4821–4826

Günthert U, Hofmann M, Rudy W, Reber S, Zöller M, Haußmann I, Matzku S, Wenzel A, Ponta H, Herrlich P (1991) A new variant of glycoprotein CD44 confers metastatic potential to rat carcinoma cells. Cell 65:13–24

Guo Y, Ma J, Wang J, Che X, Narula J, Bigby M, Wu M, Sy M-S (1994) Inhibition of human melanoma growth and metastasis in vivo by anti-CD44 monoclonal antibody. Cancer Res 54:1561–1565

Hamann A, Andrew DP, Jablonski-Westrich D, Holzmann B, Butcher EC (1994) Role of α_4-integrins in lymphocyte homing to mucosal tissues in vivo. J Immunol 152:3282–3293

Hardingham TE, Fosang AJ (1992) Proteoglycans: many forms and many functions. FASEB J 6:861–870

Harning R, Myers C, Merluzzi VJ (1993) Monoclonal antibodies to lymphocyte function-associated antigen-1 inhibit invasion of human lymphoma and metastasis of murine lymphoma. Clin Exp Metastasis 11:337–342

Hu MC-T, Crowe DT, Weissman IL, Holzmann B (1992) Cloning and expression of mouse integrin β_p (β_7): A functional role in Peyer's patch-specific lymphocyte homing. Proc Natl Acad Sci USA 89:8254–8258

Huang Y-W, Richardson JA, Vitetta ES (1995) Anti-CD54 (ICAM-1) has antitumor activity in SCID mice with human myeloma cells. Cancer Res 55:610–616

Jalkanen S, Bargatze RF, de los Toyos J, Butcher EC (1987) Lymphocyte recognition of high endothelium: antibodies to distinct epitopes of an 85–95-kD glycoprotein antigen differentially inhibit lymphocyte binding to lymph node, mucosal, or synovial endothelial cells. J Cell Biol 105:983–990

Johnston SC, Dustin ML, Hibbs ML, Springer TA (1990) On the species specificity of the interaction of LFA-1 with intercellular adhesion molecules. J Immunol 145:1181–1187

Katoh S, Zheng Z, Oritani K, Shimozato T, Kincade PW (1995) Glycosylation of CD44 negatively regulates its recognition of hyaluronan. J Exp Med 182:419–429

Koopman G, van Kooyk Y, de Graaff M, Meyer CJLM, Figdor CG, Pals ST (1990) Triggering of the CD44 antigen on T lymphocytes promotes T cell adhesion through the LFA-1 pathway. J Immunol 145:3589–3593

Koopman G, Heider K-H, Horst E, Adolf GR, van den Berg F, Ponta H, Herrlich P, Pals ST (1993) Activated human lymphocytes and aggressive non-Hodgkin's lymphoma express a homologue of the rat metastasis-associated variant of CD44. J Exp Med 177:897–904

Laurent TC, Fraser JRE (1992) Hyaluronan. FASEB J 6:2397–2404

Lesley J, Hyman R, Kincade PW (1993) CD44 and its interaction with extracellular matrix. Adv Immunol 54:271–335

Lesley J, English N, Perschl A, Gregoroff J, Hyman R (1995) Variant cell lines selected for alterations in the function of the hyaluronan receptor CD44 show differences in glycosylation. J Exp Med 182: 431–437

Ley K, Tedder TF (1995) Leukocyte interactions with vascular endothelium. New insights into selectin-mediated attachment and rolling. J Immunol 155:525–528

Liao H-X, Levesque MC, Patton K, Bergamo B, Jones D, Moody MA, Telen MJ, Haynes BF (1993) Regulation of human CD44H and CD44E isoform binding to hyaluronan by phorbol myristate acetate and anti-CD44 monoclonal and polyclonal antibodies. J Immunol 151:6490–6499

Lokeshwar VB, Fregien N, Bourguignon LYW (1994) Ankyrin-binding domain of CD44 (GP85) is required for the expression of hyaluronic acid-mediated adhesion function. J Cell Biol 126:1099–1109

Lugasi H, Hajos S, Murphy JR, Strom TB, Nichols J, Peñarroja C, Naor D (1990) Murine spontaneous T-cell leukemia constitutively expressing IL-2 receptor-A model for human T-cell malignancies expressing IL-2 receptor. Int J Cancer 45:163–167

Mackay CR (1992) Migration pathways and immunologic memory among T lymphocytes. Semin Immunol 4:51–58

Mackay CR, Marston WL, Dudler L (1990) Naive and memory T cells show distinct pathways of lymphocyte recirculation. J Exp Med 171:801–817

Matsumura Y, Sugiyama M, Matsumura S, Hayle AJ, Robinson P, Smith JC, Tarin D (1995) Unusual retention of introns in CD44 gene transcripts in bladder cancer provides new diagnostic and clinical oncological opportunities. J Pathol 177:11–20

Murakami S, Shimabukuro Y, Miki Y, Saho T, Hino E, Kasai D, Nozaki T, Kusumoto Y, Okada H (1994) Inducible binding of human lymphocytes to hyaluronate via CD44 does not require cytoskeleton association but does require new protein synthesis. J Immunol 152:467–477

Naor D, Vogt Sionov R, Ish-Shalom D (1997) CD44: structure, function and association with the malignant process. Adv Cancer Res 71:241–319

Pals ST, Hogervorst F, Keizer GD, Thepen T, Horst E, Figdor CC (1989) Identification of a widely distributed 90-kDa glycoprotein that is homologous to the Hermes-1 human lymphocyte homing receptor. J Immunol 143:851–857

Parkhurst MR, Saltzman WM (1992) Quantification of human neutrophil motility in three-dimensional collagen gels. Effect of collagen concentration. Biophys J 61:306–315

Patel DD, Hale LP, Whichard LP, Radcliff G, Mackay CR, Haynes BF (1995) Expression of CD44 molecules and CD44 ligands during human thymic fetal development: expression of CD44 isoforms is developmentally regulated. Int Immunol 7:277–286

Peach RJ, Hollenbaugh D, Stamenkovic I, Aruffo A (1993) Identification of hyaluronic acid binding sites in the extracellular domain of CD44. J Cell Biol 122:257–264

Perschl A, Lesley J, English N, Trowbridge I, Hyman R (1995) Role of CD44 cytoplasmic domain in hyaluronan binding. Eur J Immunol 25:495–501

Picker LJ, Butcher EC (1992) Physiological and molecular mechanisms of lymphocyte homing. Annu Rev Immunol 10:561–591

Pillemer G, Lugasi-Evgi H, Scharovsky G, Naor D (1992) Insulin dependence of murine lymphoid T-cell leukemia. Int J Cancer 50:80–85

Qian F, Vaux DL, Weissman IL (1994) Expression of the integrin α4β1 on melanoma cells can inhibit the invasive stage of metastasis formation. Cell 77:335–347

Rincon J, Prieto J, Patarroyo M (1992) Expression of integrins and other adhesion molecules in Epstein-Barr virus-transformed B lymphoblastoid cells and Burkitt's lymphoma cells. Int J Cancer 51:452–458

Ristamäki R, Joensuu H, Söderström K-O, Jalkanen S (1995) CD44v6 expression in non-Hodgkin's lymphoma: an association with low histological grade and poor prognosis. J Pathol 176:259–267

Rocha M, Krüger A, Umansky V, von Hoegen P, Naor D, Schirrmacher V (1996) Dynamic expression changes in vivo of adhesion and costimulatory molecules determine load and pattern of lymphoma liver metastasis. Clin Cancer Res 2:811–820

Rodrigues M, Nussenzweig RS, Romero P, Zavala F (1992) The in vivo cytotoxic activity of $CD8^+$ T cell clones correlates with their levels of expression of adhesion molecules. J Exp Med 175:895–905

Rudy W, Hofmann M, Schwartz-Albiez R, Zöller M, Heider K-H, Ponta H, Herrlich P (1993) The two major CD44 proteins expressed on a metastatic rat tumor cell line are derived from different splice variants: each one individually suffices to confer metastatic behavior. Cancer Res 53:1262–1268

Ruggiero RA, Bustuoabad OD, Bonfil RD, Meiss RP, Pasqualini CD (1985) "Concomitant immunity" in murine tumours of non-detectable immunogenicity. Br J Cancer 51:37–48

Ruoslahti E (1991) Integrins. J Clin Invest 87:1–5

Sanchez-Madrid F, Simon P, Thompson S, Springer TA (1983) Mapping of antigenic and functional epitopes on the α- and β-subunits of two related mouse glycoproteins involved in cell interactions, LFA-1 and MAC-1. J Exp Med 158:586–602

Screaton GR, Bell MV, Jackson DG, Cornelis FB, Gerth U, Bell JI (1992) Genomic structure of DNA encoding the lymphocyte homing receptor CD44 reveals at least 12 alternatively spliced exons. Proc Natl Acad Sci USA 89:12160–12164

Seiter S, Arch R, Reber S, Komitowski D, Hofmann M, Ponta H, Herrlich P, Matzku S, Zöller M (1993) Prevention of tumor metastasis formation by anti-variant CD44. J Exp Med 177:443–455

Sharon R, Pillemer G, Ish-Shalom D, Kalman R, Ziv E, Berry EM, Naor D (1993) Insulin dependence of murine T-cell lymphoma. II. Insulin-deficient diabetic mice and mice fed low-energy diet develop resistance to lymphoma growth. Int J Cancer 53:843–849

Sleeman JP, Arming S, Moll JF, Hekele A, Rudy W, Sherman LS, Kreil G, Ponta H, Herrlich P (1996) Hyaluronate-independent metastatic behavior of CD44 variant-expressing pancreatic carcinoma cells. Cancer Res 56:3134–3141

Springer TA (1994) Traffic signals for lymphocyte recirculation and leukocyte emigration: the multistep paradigm. Cell 76:301–314

Stauder R, Eisterer W, Thaler J, Günthert U (1995) CD44 variant isoforms in non-Hodgkin's lymphoma: a new independent prognostic factor. Blood 85:2885–2899

Strauch UG, Lifka A, Goßlar U, Kilshaw PJ, Clements J, Holzmann B (1994) Distinct binding specificities of integrins α4β7 (LPAM-1), α4β1 (VLA-4), and αIELβ7. Int Immunol 6:263–275

Sugimoto K, Tsurumaki Y, Hoshi H, Kadowaki S, Le Bousse-Kerdiles MC, Smadja-Joffe F, Mori KJ (1994) Effects of anti-CD44 monoclonal antibody on adhesion of erythroid leukemic cells (ELM-I-1) to hematopoietic supportive cells (MS-5): CD44, but not hyaluronate-mediated, cell-cell adhesion. Exp Hematol 22:488–494

Sy M-S, Guo Y-J, Stamenkovic I (1991) Distinct effects of two CD44 isoforms on tumor growth in vivo. J Exp Med 174:859–866

Terpe H-J, Koopmann R, Imhof BA, Günthert U (1994) Expression of integrins and CD44 isoforms in non-Hodgkin's lymphomas: CD44 variant isoforms are preferentially expressed in high-grade malignant lymphomas. J Pathol 174:89–100

Thomas L, Byers HR, Vink J, Stamenkovic I (1992) CD44H regulates tumor cell migration on hyaluronate-coated substrate. J Cell Biol 118:971–977

Toole BP, Biswas C, Gross J (1979) Hyaluronate and invasiveness of the rabbit V2 carcinoma. Proc Natl Acad Sci USA 76:6299–6303

Toyama-Sorimachi N, Miyake K, Miyasaka M (1993) Activation of CD44 induces ICAM-1/LFA-1-independent, Ca^{2+}, Mg^{2+}-independent adhesion pathway in lymphocyte-endothelial cell interaction. Eur J Immunol 23:439–446

Uff CR, Neame SJ, Isacke CM (1995) Hyaluronan binding by CD44 is regulated by a phosphorylation-independent mechanism. Eur J Immunol 25:1883–1887

van Weering DHJ, Baas PD, Bos JL (1993) A PCR-based method for the analysis of human CD44 splice products. PCR Methods Appl 3:100–106

Vogt Sionov R, Naor D (1997) Hyaluronan-independent lodgment of $CD44^+$ lymphoma cells in lymphoid organs. Int J Cancer 71:462–469

Walter J, Schirrmacher V, Mosier D (1995) Induction of CD44 expression by the Epstein-Barr virus latent membrane protein LMP1 is associated with lymphoma dissemination. Int J Cancer 61:363–369

Wielenga VJM, Heider K-H, Offerhaus GJA, Adolf GR, van den Berg FM, Ponta H, Herrlich P, Pals ST (1993) Expression of CD44 variant proteins in human colorectal cancer is related to tumor progression. Cancer Res 53:4754–4756

Wu L, Kincade PW, Shortman K (1993) The CD44 expressed on the earliest intrathymic precursor population functions as a thymus homing molecule but does not bind to hyaluronate. Immunol Lett 38:69–75

Yamada KM (1991) Adhesive recognition sequences. J Biol Chem 266:12809–12812

Yeatman TJ, Nicolson GL (1993) Molecular basis of tumor progression: mechanisms of organ-specific tumor metastasis. Semin Surg Oncol 9:256–263

Zahalka MA, Naor D (1994) β2-integrin dependent aggregate formation between LB T cell lymphoma and spleen cells: assessment of correlation with spleen invasiveness. Int Immunol 6:917–924

Zahalka MA, Okon E, Naor D (1993) Blocking lymphoma invasiveness with a monoclonal antibody directed against the β-chain of the leukocyte adhesion molecule (CD18). J Immunol 150:4466–4477

Zahalka MA, Okon E, Gosslar U, Holzmann B, Naor D (1995) Lymph node (but not spleen) invasion by murine lymphoma is both CD44- and hyaluronate-dependent. J Immunol 154:5345–5355

Zhang L, Underhill CB, Chen L (1995) Hyaluronan on the surface of tumor cells is correlated with metastatic behavior. Cancer Res 55:428–433

Zheng Z, Katoh S, He Q, Oritani K, Miyake K, Lesley J, Hyman R, Hamik A, Parkhouse RME, Farr AG, Kincade PW (1995) Monoclonal antibodies to CD44 and their influence on hyaluronan recognition. J Cell Biol 130:485–495

Zhou DFH, Ding JF, Picker LJ, Bargatze RF, Butcher EC, Goeddel DV (1989) Molecular cloning and expression of Pgp-1. The mouse homolog of the human H-CAM (Hermes) lymphocyte homing receptor. J Immunol 143:3390–3395

The Ups and Downs of $\alpha_2\beta_1$-Integrin Expression: Contributions to Epithelial Cell Differentiation and the Malignant Phenotype

M.M. ZUTTER and S.A. SANTORO

1 The Integrin Family of Cell Adhesion Receptors 167
2 Expression of $\alpha_2\beta_1$-Integrin Is Associated with Orderly, Regulated Epithelial Differentiation 168
3 Integrin Expression in Human Malignancy 169
4 Alterations in Integrin Expression May Correlate With Metastatic Potential 170
5 The Role of $\alpha_2\beta_1$-Integrin in Epithelial Cells: Experimental Approaches 171
5.1 A Gain of Function Model 172
5.2 A Loss of Function Model 174
6 The Role of $\alpha_2\beta_1$-Integrin in Mesenchymal Cells Differs Dramatically From the Role in Epithelial Cells. 179
7 The Molecular Mechanisms That Regulate $\alpha_2\beta_1$-Integrin Expression 180
8 The $\alpha_2\beta_1$-Integrin Is Required for Normal Glandular Differentiation and May Function as a Tumor Suppressor Gene 182
References 182

1 The Integrin Family of Cell Adhesion Receptors

The integrins are a family of cell surface adhesion receptors that mediate adhesion to either components of the extracellular matrix or to other cells. The integrins are noncovalently associated, heterodimeric glycoproteins composed of distinct α and β subunits of which at least 14 α and nine β subunits have been identified (RUOSLAHTI 1991; HYNES 1992; ALBELDA 1993). The β_1 family of integrins represent the major class of cell substrate receptors with specificities primarily for collagens, laminins, and fibronectins. Ligand specificity is a function of the particular α-β combination with a great deal of apparent redundancy within the system. For example, many integrins may bind a given extracellular matrix molecule and a single integrin may bind more than one matrix molecule. Recent evidence from several laboratories suggests that some of the apparent redundancy observed at the level of adhesion is

Department of Pathology, Division of Anatomic Pathology, Barnes Hospital, 1 Barnes Plaza, Washington University School of Medicine, St. Louis, MO 63110, USA not at all redundancy since each receptor may mediate distinct post-receptor occupancy events such as cell differentiation (DEDHAR et al. 1987; REICHARDT and TOMASELLI 1991), alteration in gene expression (WERB et al. 1989; DAMSKY and WERB 1992), ion fluxes through membrane channels (SCHWARTZ and DENNINGHOFF 1994), and regulation of tumor progression, invasion and metastasis.

One of the earliest suggestions that integrins play a role in differentiation and malignancy came from studies of the malignant transformation of cells in culture. PLANTEFABER and HYNES (1989) demonstrated that oncogenic transformation of rodent fibroblasts with Rous sarcoma virus encoding the *src* oncogene or murine sarcoma virus encoding the *ras* oncogene led to reduced expression of the $\alpha_5\beta_1$-integrin and two other unidentified integrins. Expression of the $\alpha_3\beta_1$-integrin was retained. Later, DEDHAR and SAULNIER (1990) demonstrated that treatment of a human osteogenic sarcoma cell line (HOS) with *N*-methyl-*N'*-nitro-*N*-nitrosoguanidine (MNNG), a potent carcinogen, altered integrin expression. Increased expression of the $\alpha_6\beta_1$-, $\alpha_2\beta_1$-, and $\alpha_1\beta_1$-integrins contrasted with reduction in $\alpha_v\beta_3$ expression and no change in $\alpha_5\beta_1$ or $\alpha_3\beta_1$ expression. The MNNG-treated cells exhibited greater invasiveness in an in vitro model that could be blocked by monoclonal antibodies directed against the β_1-integrin, suggesting that the β_1-integrin family plays a major role in the invasive potential of sarcomatous cells. In both of these examples, transformation was associated with morphologic alterations and increased invasiveness, suggesting that changes in integrin receptors might contribute to changes in cell phenotype associated with malignant transformation.

2 Expression of $\alpha_2\beta_1$-Integrin Is Associated with Orderly, Regulated Epithelial Differentiation

In recent years, our laboratory has focused on the mechanisms and functions of cellular adhesion to collagens. As a result of studies carried out in our laboratory, and in the laboratories of our colleagues, the role of the $\alpha_2\beta_1$-integrin as a collagen receptor, initially on platelets and subsequently on other cell types, has been clearly established (SANTORO and ZUTTER 1995). On some cell types, a more activated form of the receptor may serve as both a collagen and as a laminin receptor (ELICES and HEMLER 1989; LANGUINO et al. 1989; KIRCHOFER et al. 1990). As the function of the $\alpha_2\beta_1$-integrin was being elucidated, the identity of the integrin with the platelet membrane Ia-IIa complex, very late activation antigen 2 (VLA-2) on activated T cells and the ECMR II (class 2 extracellular matrix receptor) on fibroblasts was also established. These observations suggested that the adhesive mechanism mediated by the $\alpha_2\beta_1$-integrin was employed by cells other than blood platelets. This recognition caused us to undertake an extensive study of the tissue distribution of the $\alpha_2\beta_1$-integrin (ZUTTER and SANTORO 1990). We observed that the receptor was widely distributed. In addition to its expression by fibroblasts and endothelial cells, high level expression was observed on numerous epithelial cells including keratinizing and non-keratinizing stratified squamous epithelium, ciliated columnar epithelium of the respiratory tract, the epithelial cells of the gastrointestinal and urinary tract, and the glandular epithelium of the breast. A recurrent finding in our study was the association of increased expression of the $\alpha_2\beta_1$-integrin with the orderly regulated proliferation of epithelial cells.

3 Integrin Expression in Human Malignancy

These observations prompted an investigation of $\alpha_2\beta_1$-integrin expression in malignancy using breast cancer as a model (ZUTTER et al. 1990). This initial consideration of integrin expression in human solid tumors revealed that the $\alpha_2\beta_1$-integrin was highly expressed in the epithelium of ducts and ductules of normal breast tissue. Normal or close to normal levels of expression were observed in benign lesions such as fibroadenomas or papillomas. In contrast, markedly reduced or undetectable $\alpha_2\beta_1$ expression was seen in poorly differentiated adenocarcinomas. Well differentiated adenocarcinomas exhibited intermediate levels of expression. Similar but less extensive changes were observed for the $\alpha_5\beta_1$ (fibronectin receptor) and the $\alpha_v\beta_3$- (vitronectin receptor) integrins. Significant residual expression of the β_1 subunit on poorly differentiated tumors suggested that the expression of other members of the β_1 family was not reduced to the same low levels as the α_2 subunit.

To extend our immunohistochemical observations and explore the regulation of altered α_2-integrin expression in breast cancer, we employed in situ hybridization to assay the levels of integrin mRNA in the same panel of tumors we had previously studied by immunohistochemistry (ZUTTER et al. 1993). Normal breast ducts and ductules expressed high steady state levels of α_2 mRNA detected by an antisense α_2 cDNA probe. The level of α_2 mRNA was slightly decreased in well differentiated lesions and was more significantly decreased in moderately differentiated lesions. In poorly differentiated tumors, steady state levels of α_2 mRNA were markedly decreased but still detectable. This study, which represented an initial analysis of integrin gene expression in cancer at the molecular level documented that decreased integrin protein expression is a consequence of altered α_2-integrin gene expression. These studies also established that expression of the $\alpha_2\beta_1$-integrin at both the protein and mRNA levels is decreased in adenocarcinoma of the breast in a manner that correlates with the loss of tumor cell differentiation (ZUTTER et al. 1990, 1993). Subsequent studies have addressed the role of the $\alpha_2\beta_1$-integrin expression in epithelial differentiation and the consequences of its altered expression in cancer progression, as well as the molecular mechanisms by which expression of the α_2-integrin subunit is regulated (see below).

Studies by other investigators (KOUKOULIS et al. 1991; PIGNATELLI et al. 1991a, 1992) confirmed our key observation that $\alpha_2\beta_1$-integrin expression is decreased in adenocarcinoma of the breast in a manner that correlates with the degree of tumor cell differentiation. Studies of other adenocarcinomas (i.e., colon, prostate, lung, pancreas, and skin) have yielded similar findings regarding $\alpha_2\beta_1$ expression (KORETZ et al. 1991; STALLMACH et al. 1992; HALL et al. 1991; BONKOFF et al. 1993; STAMP and PIGNATELLI 1991). Studies of colorectal cancer have shown that $\alpha_2\beta_1$ is consistently lost or diminished in moderately and poorly differentiated colorectal carcinoma in a manner similar to that described for breast cancer (PIGNATELLI 1990; PIGNATELLI et al. 1991b; KORETZ et al. 1991; KOUKOULIS et al. 1993). In addition, diminution or loss of $\alpha_2\beta_1$-integrin by renal, pancreatic, and lung cancer was associated with poorly differentiated lesions (KORHONEN et al. 1992; WEINEL et al. 1992; DAMJANOVICH et al. 1992). Review of the existing literature reveals that decreased expression of $\alpha_2\beta_1$ is the most common change in integrin expression in epithelial malignancies (ALBELDA 1993; VARNER and CHERESH 1996).

Alterations of expression or cellular localization of other integrins have been described in carcinoma of the breast and other epithelial malignancies. The $\alpha_3\beta_1$ and the $\alpha_6\beta_4$-integrins are expressed at high levels in most normal epithelial cells and associated with the differentiated epithelial phenotype. Reports of reduced α_3-integrin subunit expression in breast, colorectal, and pancreatic carcinomas have appeared (STALLMACH et al. 1992; WEINEL et al. 1992). Correlation of the α_3-integrin expression with the state of differentiation has been less well substantiated. The literature regarding changes in the α_6-integrin subunit expression is inconsistent. In contrast to the α_2 and α_3 subunits which associate with only the β_1 subunit, the α_6 subunit can pair with either the β_1 or β_4 subunits introducing an additional level of complexity (HEMLER et al. 1989; KAJIJI et al. 1989). Alteration of the pattern of heterodimerization might not be reflected in an overall change in the α_6 subunit expression (SONNENBERG et al. 1990). Changes in $\alpha_6\beta_4$ expression appear more variable than the rather consistent diminution or loss of $\alpha_2\beta_1$ expression by most carcinomas (KENNEL et al. 1986; LIEBERT et al. 1993; KOUKOULIS et al. 1991, 1993; PIGNATELLI et al. 1992). In contrast to integrins that have been primarily associated with the state of epithelial differentiation, expression of the $\alpha_5\beta_1$- and $\alpha_v\beta_3$-integrins have been reported to be increased, decreased, or unchanged in adenocarcinoma of the breast (VARNER and CHERESH 1996).

4 Alterations in Integrin Expression May Correlate With Metastatic Potential

As described above, the association of high level $\alpha_2\beta_1$-integrin expression with normal epithelial differentiation and the loss of $\alpha_2\beta_1$ expression with concomitant loss of glandular differentiation has now been well established in breast and other epithelial tumors. These findings suggested that poorly differentiated $\alpha_2\beta_1$-integrin negative tumors might be more invasive and metastatic in vivo. However, a direct correlation between integrin expression and the invasive and metastatic potential of epithelial malignancies has been difficult to establish. In one clinical study that compared the expression of integrin subunits by 12 benign and 61 malignant breast samples, expression of the $\alpha_1\beta_1$, $\alpha_2\beta_1$, $\alpha_3\beta_1$, $\alpha_6\beta_1$, $\alpha_v\beta_1$, and $\alpha_v\beta_5$ integrins was significantly reduced in the malignant tissue (GUI et al. 1995a). The loss of integrin subunit expression correlated with the presence of axillary lymph node metastases. In fact, 70% of the patients who had positive lymph node metastases had no detectable expression of β_1-integrins. In a multivariant analysis to determine whether integrin expression was independent of other predictors of axillary node metastases, loss of expression of either the α_1, α_v, or β_1 subunit was found to be an important independent predictor of axillary spread. In a related study by the same group, the adhesive properties of breast cancer cells from axillary node negative specimens and axillary node positive specimens were examined (GUI et al. 1995b). Primary breast tumor cells from women without axillary node metastases adhered significantly better to laminin than did primary tumor cells derived from women with axillary lymph node metastases. Adhesion to laminin was inhibited by the anti-α_1-integrin monoclonal antibody, P1E6, or by the inhibitory anti-β_1-integrin subunit monoclonal antibody, mab13. The ability of cells to adhere to laminin directly correlated with the expression of α_2- and β_1-integrin subunits by immunohistochemistry. Additional studies have correlated integrin expression with tumor progression in colorectal cancer. LINDMARK et al. (1993) demonstrated that the loss in basolateral expression of both the $\alpha_2\beta_1$-and $\alpha_3\beta_1$-integrins related directly to tumor cell differentiation, advanced Dukes stage and poor survival. These studies suggest that in both breast and colon cancer the extent of altered expression and function of the $\alpha_2\beta_1$ and $\alpha_3\beta_1$ integrins may correlate with prognosis.

5 The Role of $\alpha_2\beta_1$-Integrin in Epithelial Cells: Experimental Approaches

To date, few experimental studies had addressed the role of altered integrin expression in epithelial differentiation or the potential role of altered integrin expression in tumor cell invasion and metastasis. Peptides containing the sequence RGD, a recognition sequence for some integrins, but not for the $\alpha_2\beta_1$-integrin, have been shown to inhibit the pulmonary metastasis of B16-F10 melanoma cells injected into the tail veins of mice (HUMPHRIES et al. 1986). GIANCOTTI and RUOSLAHTI (1990) demonstrated that overexpression of the $\alpha_5\beta_1$-integrin, a fibronectin receptor, by Chinese hamster ovary cells resulted in a loss of tumorigenicity. Overexpression of $\alpha_5\beta_1$ induced increased fibronectin matrix formation, decreased cell migration, decreased saturation density in culture and reduced tumor formation in nude mice. These findings suggest that $\alpha_5\beta_1$ is involved in extracellular matrix deposition and that expression of the receptor serves to maintain a differentiated phenotype, to prevent cell proliferation and tumor cell invasion in the Chinese hamster ovary cell model. A role for $\alpha_5\beta_1$ in tumorigenesis was also demonstrated by SCHREINER et al. (1991) who isolated naturally occurring clonal lines of Chinese hamster ovary cells expressing increased or decreased levels of the $\alpha_5\beta_1$-integrin. Clones expressing lower levels of $\alpha_5\beta_1$ produced larger tumors following subcutaneous injection and grew more rapidly than did control cells.

The dramatic and consistent alterations in expression of the $\alpha_2\beta_1$-integrin in epithelial malignancy and the correlation between the loss of $\alpha_2\beta_1$-integrin expression and the differentiated epithelial phenotype raised critical questions: What is the role of α_2-integrin expression in the maintenance of the differentiated epithelial phenotype and what is the contribution of altered expression to the malignant behavior of breast cancer cells? To begin to address these questions, we developed two complimentary approaches; a gain of function model and a loss of function model. We exploited both approaches to examine directly the effects of $\alpha_2\beta_1$-integrin expression by mammary epithelial cells.

5.1 A Gain of Function Model

For the gain of function model, we identified a breast cancer cell line, the Mm5MT cell line, a mouse mammary tumor virus (MMTV)-induced tumorigenic murine breast cancer cell line that expressed no detectable α_2-integrin protein or mRNA. A full-length α_2-integrin cDNA was introduced into the Mm5MT cell line and six clonal cell lines expressing similar levels of α_2-integrin protein and mRNA were identified (Zutter et al. 1995a).

The most striking initial difference between the α_2 transfectants and the parental or control transfectants was their morphologic appearance in culture. The parental Mm5MT cell line grew as spindle-shaped clusters that rapidly formed large multilayered colonies with poorly defined margins. Growth was not contact inhibited. In contrast, the α_2 transfectants grew as a monolayer of polygonal cells with a cobblestone appearance and were contact inhibited. The restoration of contact inhibition suggested that expression of the $\alpha_2\beta_1$-integrin by this poorly differentiated breast carcinoma altered the growth characteristics and the transformed phenotype of the parental cell line.

Interestingly, adhesion to collagen was comparable for controls and α_2 transfectants. The control Mm5MT cells and α_2 transfectants expressed the α_1-integrin subunit at high and equivalent levels. The adhesion to collagen by parental Mm5MT cell was likely mediated by the $\alpha_1\beta_1$-integrin, which also can serve as a collagen/laminin receptor, and which was expressed at a high level by parental and control cells (Ignatius and Reichardt 1988; Ignatius et al. 1990; Tawil et al. 1990; Kern et al. 1993). Expression of the $\alpha_2\beta_1$-integrin did not alter the level of $\alpha_1\beta_1$-integrin expression and did not alter the adhesion of the cells to collagen.

Although the ability of the α_2-integrin-expressing clones to adhere to collagen was similar to the Mm5MT parental cells, the ability of the clones to spread on type I collagen was markedly altered. The parental Mm5MT and the control Mm5MT-neo cells spread and developed elongate processes on type I collagen substrates. In contrast, the α_2-expressing clones adhered but spread slowly and formed only short rudimentary processes. The difference in cell spreading following adhesion to collagen suggests that events occurring after interaction of the $\alpha_2\beta_1$-integrin with the matrix can modify the signaling events mediated by other collagen receptors present on the Mm5MT cells. The α_2 transfectants migrated more slowly toward a gradient of type I collagen in a haptotaxis assay. Invasion of the α_2-transfected Mm5MT cells through a gel of reconstituted basement membrane (Matrigel) was also markedly reduced when compared to parental Mm5MT and controls.

We compared the rate of cell proliferation and the saturation density of the α_2-expressing clones to that of parental Mm5MT or control transfectants. Expression of the α_2-integrin subunit did not affect the growth rate per se, but did profoundly alter the saturation density. The lack of effect of α_2-integrin expression on proliferation stands in sharp contrast to the marked influence of $\alpha_5\beta_1$-integrin expression on the rate of cell proliferation (GIANCOTTI and RUOSLAHTI 1990). Re-expression of the $\alpha_2\beta_1$-integrin did alter anchorage independent growth in soft agar. The Mm5MT and Mm5MT-neo controls formed abundant large colonies (greater than 0.5 mm diameter) over an 18 day period; the α_2-transfected cell lines formed only few (3–4) colonies over the same time period. Thus, although the expression of the α_2-integrin subunit did not affect growth rate of cells attached to a substrate, α_2-integrin expression dramatically reduced the capacity for anchorage independent growth.

The inability of the α_2-integrin-expressing clones to form colonies in soft agar suggested that the tumorigenic potential of the α_2-expressing clones might also be altered in vivo. Subcutaneous injection of 2.5×10^6 Mm5MT or Mm5MT-neo cells into the flank of weanling severe combined immunodeficiency (SCID) mice resulted in large tumors ($>1.5\ cm^3$ in volume) in all ten mice within a 3 week interval. In some animals, tumors grew as large as 3.5 cm in greatest diameter (Table 1). In contrast, the α_2-expressing clones either completely failed to form identifiable tumors or formed only small, elongated, single tumors ($<0.07\ cm^3$ in volume) along the needle track at the site of injection (Table 1). These findings suggest that re-

Table 1. In vivo tumorigenicity

Cell line	Mouse tumor group/total		
	Large tumors[a]	Small tumors[b]	Tumors (n)
Mm5MT	5/5	–	–
Mm5MT	5/5	–	–
B1	–	5/5	–
B2	–	3/5	2/5
C3	–	–	5/5

[a] Tumors $>1.5\ cm^3$ in volume
[b] Tumors $<0.07\ cm^3$ in volume
(From ZUTTER et al. 1995a, with permission)

expression of the α_2-integrin by a poorly differentiated breast carcinoma cell profoundly suppresses tumor development in vivo.

Normal mammary gland morphogenesis requires a complex interplay between hormonal stimulants, growth factors, and the extracellular matrix (Strange et al. 1991; Reichmann et al. 1989; Taylor-Papadimitriou et al. 1993; Oka et al. 1991). Primary mammary epithelial cells and immortalized, but nontumorigenic, mammary cell lines of both mouse and human origin form glandular structures in three-dimensional collagen gels (Emerman and Pitelka 1977; Lee et al. 1985; Bissell and Hall 1987). The formation of these structures mimics ductal morphogenesis. A role for the $\alpha_2\beta_1$-integrin in gland formation had been suggested (Berdichevsky et al. 1991). We therefore analyzed the ability of the α_2-transfected cells and controls to undergo morphogenesis in either reconstituted basement membrane gels or floating collagen gels. When grown in gels composed of reconstituted basement membrane, the Mm5MT and Mm5MT-neo cells formed large, disorganized aggregates of spindle-shaped cells (Fig. 1). In contrast, the α_2-subunit expressing clones B1, B2, and C3 formed three-dimensional organized structures including alveolar-like and elongated multilayered, duct-like structures (Fig. 1). Both alveolar structures and ducts branched extensively. Rudimentary duct-like structures also formed in collagen gels, but morphogenesis required 7–8 days instead of 3–4 days.

Based upon studies in the gain of function model, we have established that reexpression of the $\alpha_2\beta_1$-integrin by a poorly differentiated breast carcinoma cell line restores epithelial differentiation and glandular morphogenesis in vitro and alters the malignant potential in vivo. It should be emphasized that in this model, $\alpha_2\beta_1$-integrin expression is restored to a malignant cell, which our studies of human tumors revealed was lost in the malignant state.

5.2 A Loss of Function Model

We have also established a complementary loss of function model using the well-differentiated, estrogen responsive breast cancer line T47-D. Flow cytometric analysis of the T47-D cells revealed that the cells express high levels of the α_2-, α_3-, and β_1-integrin subunits and absent or low levels of the α_1 and α_5 subunits. Preliminary studies with inhibitory anti-integrin monoclonal antibodies revealed that the $\alpha_2\beta_1$-integrin is the primary mediator of T47-D cell adhesion to collagen. T47-D cells cultured in three-dimensional collagen I gels for 8–10 days organized into branching tubules with alveolar-like structures similar to the structures formed by normal mammary epithelial cells (Keely et al. 1995). Inhibitory anti-α_2 antibody, which blocked adhesion to collagen, also inhibited glandular differentiation in three-dimensional collagen matrices. We exploited antisense mRNA technology to generate stable clonal lines of T47-D cells expressing antisense α_2-integrin subunit mRNA and diminished $\alpha_2\beta_1$-integrin protein on the cell surface. Ten clonal cell lines were selected in hygromycin, either 125 mg/ml (400-series) or 250 mg/ml (700-series). Cells of the 400-series expressed moderately decreased levels of $\alpha_2\beta_1$ protein

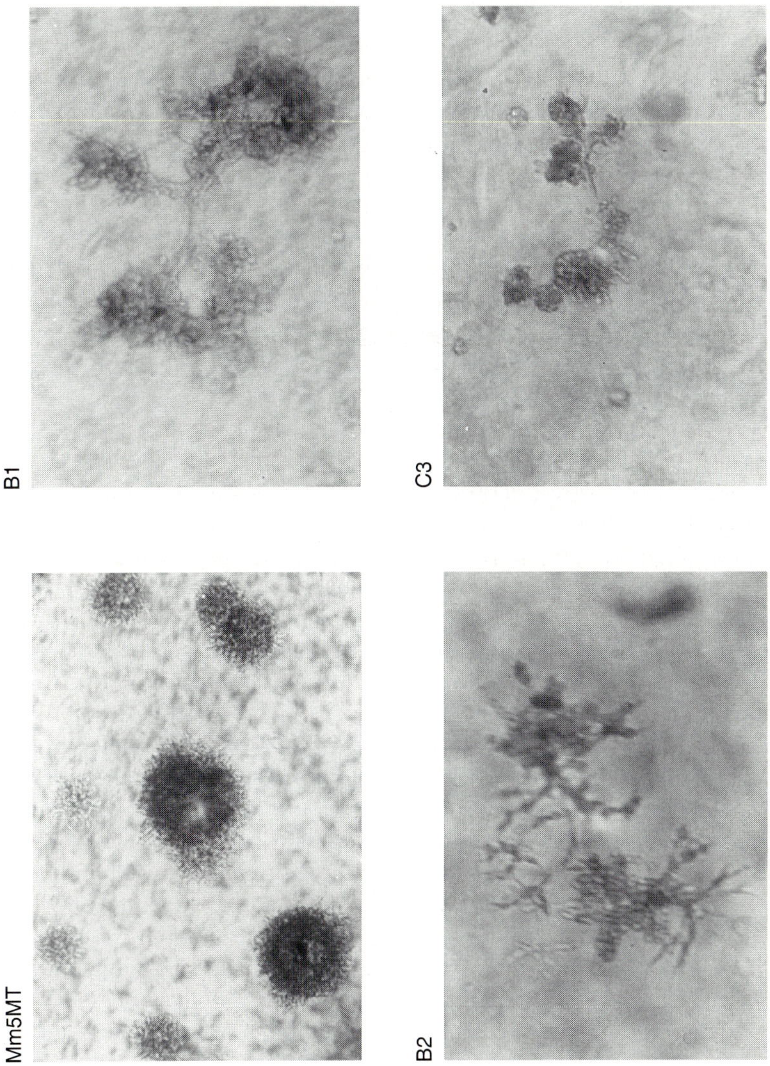

Fig. 1. Morphogenesis of Mm5MT-neo and α_2-subunit-expressing clones B1, B2, and C3 in gels of reconstituted basement membrane matrix (Matrigel). (From Zutter et al. 1995a); ×60

and cells of the 700 series expressed markedly decreased $\alpha_2\beta_1$ protein. The changes in α_2-integrin expression were specific. Expression of the α_5- and α_3-integrin subunits, for example, was unaltered.

To determine the importance of altered $\alpha_2\beta_1$ expression in the adhesive and motile properties of T47-D cells, the ability of the antisense-α_2 clones to adhere to types I and IV collagen was assayed. The α_2 antisense transfectants adhered to a lesser extent to both type I and type IV collagen than the parental or control pREP4 cells. The extent to which adhesiveness decreased correlated with decrease in $\alpha_2\beta_1$ protein expression. In contrast to the direct correlation observed between $\alpha_2\beta_1$-integrin expression and adhesion to collagen, cell motility was a more complex function of integrin expression. Cells with either normal or very low levels of $\alpha_2\beta_1$ expression exhibited little or no motility on collagen substrates (Fig. 2). In contrast, cells expressing intermediate levels of the $\alpha_2\beta_1$-integrin exhibited greater motility on collagen. Motility on fibronectin was not effected. These findings indicate that while the $\alpha_2\beta_1$-integrin is clearly required for cell migration on collagen substrates, elevated levels of $\alpha_2\beta_1$ depress migration. Migration was best supported by intermediate levels of integrin expression, a finding in accord with theoretical considerations (Lauffenberg and Horowitz 1996).

We assessed the role of $\alpha_2\beta_1$-integrin expression in the formation of glandular structures when placed in three-dimensional collagen matrices. In contrast to the ability of T47D cells to form organized structures (Fig. 3A), antisense clones with reduced $\alpha_2\beta_1$-integrin expression failed to form such structures, but grew as disorganized sheets and clumps of cells. The loss of morphologic organization was most pronounced in the 700-series (Fig. 3C), which expressed the lowest $\alpha_2\beta_1$ levels and intermediate in the 400-series (Fig. 3B), which expressed moderate levels of $\alpha_2\beta_1$-integrin. The results of these studies in the loss of function model were in accord with the results of studies employing the gain of function model. Similar studies employing antisense mRNA to inhibit $\alpha_2\beta_1$-integrin expression in renal tubular MDCK cells have also implicated the $\alpha_2\beta_1$-integrin in the branching tubulogenesis exhibited by MDCK cells in collagen gels (Saelman et al. 1995).

In summary, we have characterized in detail the role of the $\alpha_2\beta_1$-integrin in breast epithelial morphogenesis, as well as the contribution of diminished $\alpha_2\beta_1$ expression to the malignant and invasive phenotype of mammary carcinoma cells. To do so, we have developed and exploited both gain of function and loss of function models. Both approaches indicate that the $\alpha_2\beta_1$-integrin is required for epithelial differentiation and morphogenesis of breast glands and tubules and that the diminished $\alpha_2\beta_1$-integrin expression contributes to motility and the invasive behavior of tumor cells in vitro. We have also shown that reexpression of the $\alpha_2\beta_1$-integrin in a poorly differentiated, invasive breast carcinoma cell line greatly diminishes, but does not completely abrogate, the malignant potential in vivo.

Our critical observation that $\alpha_2\beta_1$-integrin expression is required for maintenance of the differentiated epithelial phenotype and glandular differentiation in vitro has been confirmed by studies from a number of our colleagues. Using a primary, human, nonmalignant, but immortalized mammary epithelial cell line, Berdichevsky et al. (1991) demonstrated that branching morphogenesis can be

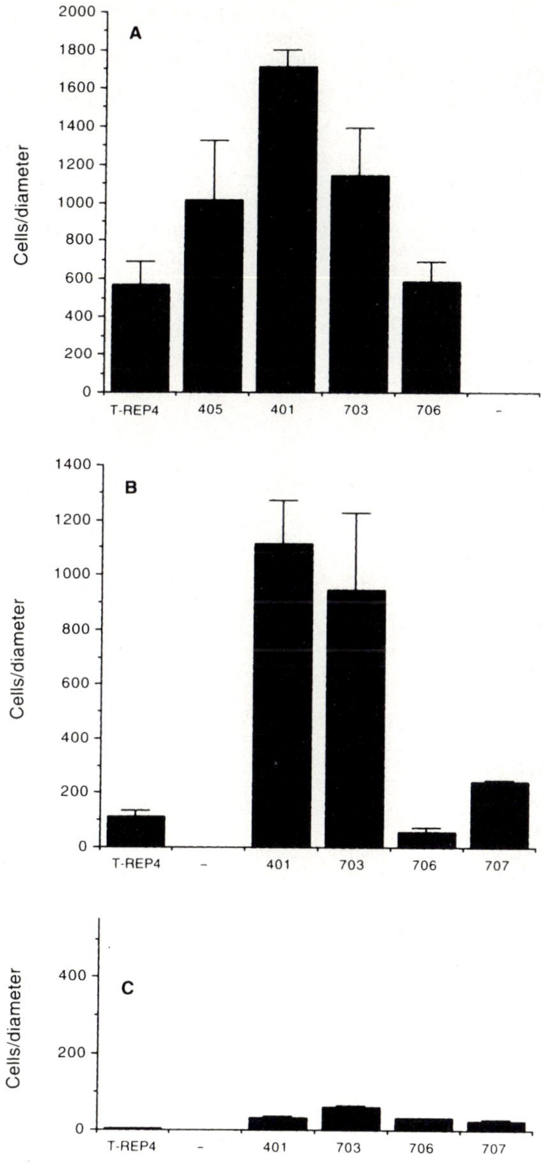

Fig. 2A–C. Cells expressing partially decreased levels of $\alpha_2\beta_1$-integrin are more motile on collagen in a haptotaxis assay. Cells were seeded into the top chamber of a Transwell and allowed to migrate for 20 h across filters coated from the underside with 30 μg/ml collagen I (**A**, **B**) or fibronectin (**C**). Two different experiments are shown for motility across collagen I. Clones are arranged from greatest adhesion to collagen (*left*) to least adhesion to collagen (*right*) and are aligned vertically for direct comparison between experiments. Cells were quantitated by counting across two diameters each of duplicate filters, and are shown ±SD. (From Keely et al. 1995)

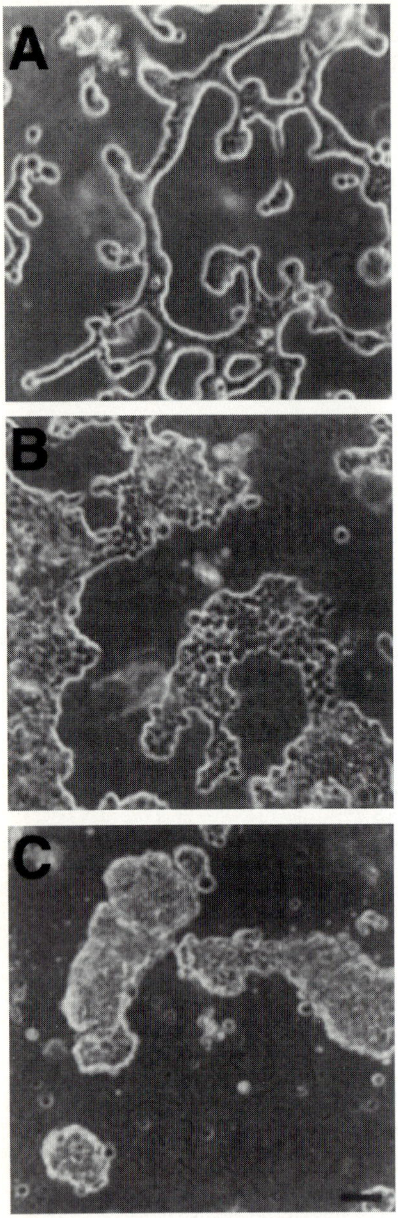

Fig. 3A–C. Antisense-expressing cells with decreased $\alpha_2\beta_1$-integrin levels exhibit disrupted morphology when cultured in three-dimensional collagen gels. Cells were cultured for 10 days in collagen gels and photographed using phase-contrast microscopy. **A** Control, T-REP4 cells formed tubule structures and organized when cultured in collagen gels. **B** Antisense-expressing clone 401 and **C** antisense-expressing clone 703 did not form tubule structures, but instead grew as disorganized sheets or clumps of cells when cultured in collagen gels. Cells were photographed using phase-contrast microscopy and are all shown at the same magnification. *Bar* in **C**, 100 mm. (From KEELY et al. 1995)

blocked with inhibitory monoclonal antibodies directed against the α_2-integrin subunit. Studies from a number of investigators suggest that the $\alpha_2\beta_1$-integrin also mediates the morphologic differentiation of colonic epithelial cells. Colorectal epithelial cells organize into glandular structures with well defined polarity in three-dimensional collagen gels (PIGNATELLI and BODMER 1988) or gels composed of basement membrane extracts (DEL BUONO et al. 1991). The ability to undergo glandular morphogenesis in three-dimensional collagen gels was inhibited by inhibitory anti-α_2 or anti-β_1 monoclonal antibodies and enhanced by TGFα induced up-regulation of $\alpha_2\beta_1$ expression on the cell surface (LIU et al. 1994). These findings support a role for the $\alpha_2\beta_1$-integrin in glandular morphogenesis of colonic epithelial cells as well as breast epithelium.

6 The Role of $\alpha_2\beta_1$-Integrin in Mesenchymal Cells Differs Dramatically From the Role in Epithelial Cells

Our studies on the role of the $\alpha_2\beta_1$-integrin in cells of epithelial origin differ from earlier findings of α_2-integrin subunit expression by cells of mesenchymal origin. CHAN et al. (1991) overexpressed the $\alpha_2\beta_1$-integrin in a rhabdomyosarcoma (RD) cell line, a skeletal muscle tumor, which did not express the $\alpha_2\beta_1$-integrin. The resulting cells were no more tumorigenic than the parental cell line, but were more metastatic when injected intravenously into nude mice. Transfection of the α_2-integrin conferred upon RD cells the ability to contract collagen matrices. SCHIRO et al. (1991) have reported that the $\alpha_2\beta_1$-integrin on fibroblasts is required for collagen gel contraction. KLEIN et al. (1991a) confirmed these observations and showed that fibroblasts increased the synthesis and transcription of the $\alpha_2\beta_1$-integrin when embedded in type I collagen gels. Gel contraction, an in vitro model for reorganization of connective tissue during wound healing, paralleled the increased expression of $\alpha_2\beta_1$ protein. Gel contraction could be blocked by inhibitory monoclonal antibodies against either the α_2-or β_1-integrin subunits. Using the same model system, melanoma cell lines were tested for their ability to contract collagen gels (KLEIN et al. 1991b). These studies suggest that the function of the $\alpha_2\beta_1$-integrin may be cell type-dependent.

The apparent contradiction between the results of studies with epithelial and other cell types in vitro is also apparent in vivo. In contrast to the diminution or loss of $\alpha_2\beta_1$-integrin expression in epithelial malignancies, expression of the $\alpha_2\beta_1$-integrin by malignant melanocytes in melanoma is markedly increased. Normal melanocytes express the α_2-integrin subunit only at very low levels. When melanocytes proliferate abnormally or become malignant, expression of the α_2-integrin subunit is aberrantly up-regulated (KLEIN et al. 1991b; DANEN et al. 1993, 1994; VAN DUINEN et al. 1994). Both benign nevi as well as cells from malignant melanoma express the $\alpha_2\beta_1$-integrin. KRAMER and MARKS (1989) initially showed that a highly metastatic melanoma cell line, MeWo, adhered tightly to type I and type IV collagens via both the $\alpha_1\beta_1$- and $\alpha_2\beta_1$-integrin receptors. Melanoma cell lines that were highly metastatic in vivo expressed high levels of the $\alpha_2\beta_1$-, as well as the $\alpha_1\beta_1$- and $\alpha_6\beta_1$-integrins. Primary nonmetastatic melanomas expressed the $\alpha_2\beta_1$-integrin at low levels (MORTARINI et al. 1991). These findings suggest that the expression and function of the $\alpha_2\beta_1$-integrin by melanocytes or mesenchymal cells is distinctly different than by an epithelial cell. The function of any specific integrin appears to be critically dependent upon the cellular environment in which it is expressed.

7 The Molecular Mechanisms That Regulate $\alpha_2\beta_1$-Integrin Expression

Recent studies have provided insight into the mechanism by which α_2-integrin subunit expression is diminished in breast cancer. Our own studies have revealed that the 5′ regulatory region of the α_2-integrin gene may be divided into three regions; a core promoter located between bp –30 and –92, a silencer located between bp –92 and –351, and a larger tissue- and differentiation-specific enhancer region located in the more distal 5′ flank (ZUTTER et al. 1994, 1995b). The core promoter is not cell type-specific and is functional in both epithelial cells and hematopoietic cells induced along the megakaryocytic pathway. The silencer strongly represses promoter activity in cells of hematopoietic lineage, but is either only weakly active or inactive as a silencer in nonhematopoietic cells, i.e., the nontumorigenic human mammary epithelial cell line MTSVI-7 and the well-differentiated epithelial cell line T47D, respectively (ZUTTER et al. 1994; YE et al. 1996). Additional enhancers in the distal 5′ flank are required for high level megakaryocytic and epithelial expression of the α_2-integrin gene. A diagram of the 5′ flank of the α_2-integrin gene from bp –3739 through exon 1 (ZUTTER et al. 1995b) demonstrates the three essential promoter/enhancer domains (Fig. 4).

We recently determined by site-directed mutagenesis of the core promoter that the two tandem Sp1 binding sites located between bp –30 and –92 are required for binding of the nuclear protein Sp1 and for promoter activity of the α_2-integrin gene (ZUTTER et al. 1997). Sp1 protein required phosphorylation for DNA-protein complex formation with the α_2-integrin core promoter region. D'SOUZA et al. (1993) showed that expression of the α_2-integrin subunit in a primary mammary epithelial cell line was reduced when the cells were transfected with the erb-B2 protooncogene. The loss of $\alpha_2\beta_1$-integrin expression by erb-B2 transfection was found to be due to a decreased steady state level of α_2-integrin mRNA resulting from reduced transcription of the α_2-integrin gene. The level of α_2 mRNA expression inversely correlated with the level of erb-B2 expression. YE et al. (1996) recently found that the reduced α_2-integrin gene expression in mammary epithelial cells transfected with erb-B2 was due, in part, to the inability of Sp1 to bind to the core promoter region, although the overall level of Sp1 protein was unchanged. These findings suggest that overexpression of erb-B2 in mammary carcinoma cells may lead to altered phosphorylation of Sp1 protein which results in an inability of Sp1 protein to bind to the core promoter of the α_2-integrin gene, thereby reducing α_2-integrin gene expression.

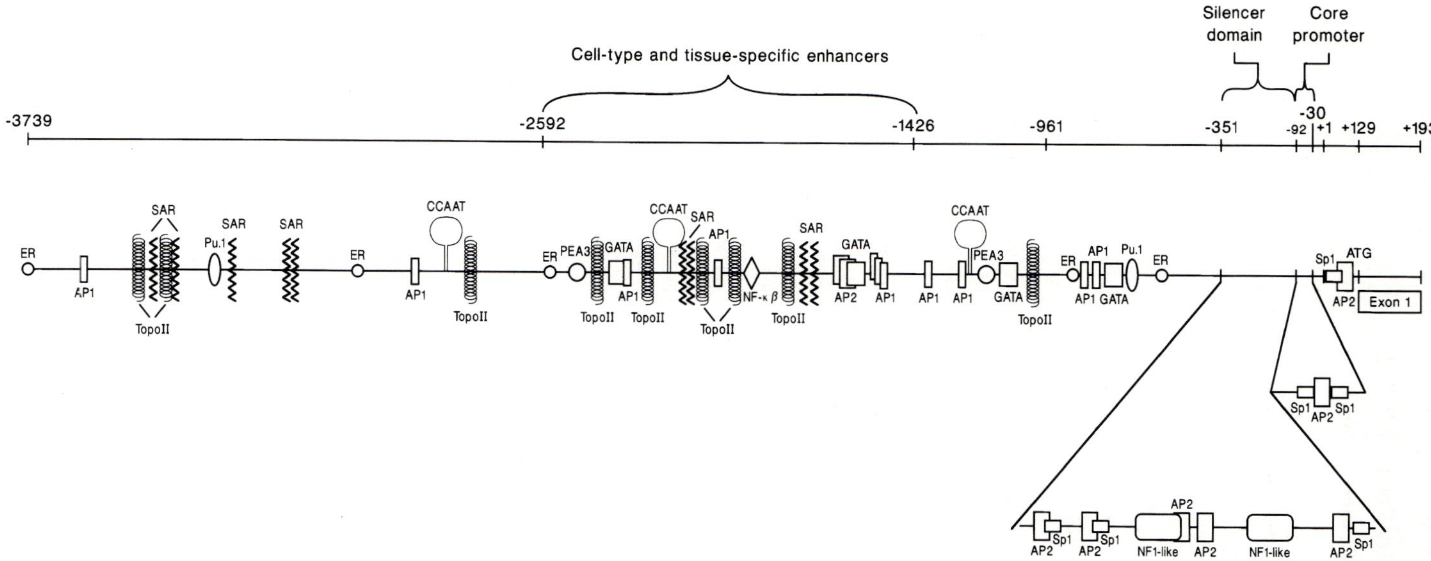

Fig. 4. The distal 5′ flank from bp –3739 through exon 1 of the α_2-integrin gene. The locations of numerous potential binding sites for ubiquitous as well as tissue-specific transcription factors are shown. In addition to numerous potential binding sites for AP1 and AP2, three perfect consensus sequences for binding to the GATA family of transcription factors are located between –961 and –2592. Additional potential binding sites include consensus sequences of NFκb, CAAT binding protein and four estrogen receptor half-sites. In addition, there are two clusters of binding sites for the enzyme topoisomerase II (*Topo II*) and scaffold-associated/matrix-associated region binding proteins (*SAR/MAR*). The complete DNA sequence of the region from –961 to –3739 has been submitted to GenBank, accession no. U31518. (From Zutter et al. 1995b)

8 The $\alpha_2\beta_1$-Integrin Is Required for Normal Glandular Differentiation and May Function as a Tumor Suppressor Gene

In summary, the function of a specific integrin is critically dependent upon the cellular environment in which it is expressed. The studies summarized in this review provide compelling evidence of the critical role for the $\alpha_2\beta_1$-integrin in normal mammary differentiation, as well as in the differentiation of other epithelial cells. It is likely that changes in $\alpha_2\beta_1$-integrin expression in breast cancer and other epithelial malignancies contribute to the altered adhesive and invasive characteristic of the tumor cells. The results of studies by our group, as well as by others, indicate that expression of the $\alpha_2\beta_1$-integrin is required for normal mammary epithelial differentiation and glandular morphogenesis. Our findings suggest that $\alpha_2\beta_1$-integrin expression diminishes the invasive and malignant phenotype of breast cancer cells and may function as a tumor suppressor for breast and other epithelial malignancies. The loss of $\alpha_2\beta_1$-integrin expression may represent one step in the progression of a normal, differentiated mammary gland epithelium to an invasive and tumorigenic breast cancer.

References

Albelda SM (1993) Biology of disease. Role of integrins and other cell adhesion molecules in tumor progression and metastasis. Lab Invest 68:4–17

Berdichevsky F, Gilbert C, Shearer M, Taylor-Papadimitriou J (1991) Collagen-induced morphogenesis of human mammary epithelial cells: the role of the $\alpha_2\beta_1$ integrin. J Cell Sci 102:437–446

Bissell MJ, Hall HG (1987) Form and function in the mammary gland: the role of extracellular matrix. In: Neville MC, Daniel CW (eds) The mammary gland: development, regulation and function. New York, Plenum, pp 97–146

Bonkoff H, Stein U, Remberger K (1993) Differential expression of α_6 and α_2 very late antigen integrins in the normal, hyperplastic and neoplastic prostate: Simultaneous demonstration of cell surface receptors and their extracellular ligands. Human Pathol 24:243–248

Chan BMC, Matsuura N, Takada Y, Zetter BR, Hemler ME (1991) In vitro and in vivo consequences of VLA-2 expression on rhabdomyosarcoma cells. Science 251:1600–1602

Damjanovich L, Albelda SM, Mette SA, Buck CA (1992) Distribution of integrin cell adhesion receptors in normal and malignant lung tissue. Am J Respir Cell Mol Biol 6:197–206

Damsky CH, Werb Z (1992) Signal transduction by integrin for extracellular matrix: cooperative processing of extracellular information. Curr Opin Cell Biol 4:772–781

Danen EH, Ten Berg PJ, Van Muijen GN, Van Hofgrootenboer AE, Bröcker EB (1994) Emergence of $\alpha_5\beta_1$ fibronectin and $\alpha_v\beta_3$ vitronectin receptor expression in melanocytic tumour progression. Histopathology 24:249–256

Danen EHJ, van Muijen GNP, van de Wiel-van Kemenade E, Jansen KFJ, Ruiter DJ, Figdor CG (1993) Regulation of integrin-mediated adhesion to laminin and collagen in human melanocytes and in non-metastatic and highly metastatic human melanoma cells. Int J Cancer 54:315–321

Dedhar S, Saulnier R (1990) Alterations in integrin receptor expression on chemically transformed human cells: specific enhancement of laminin and collagen receptor complexes. J Cell Biol 110:481–489

Dedhar S, Argraves WS, Suzuki S, Ruoslahti E, Pierschbacher MD (1987) Human osteosarcoma cells resistant to detachment by an Arg-Gly-Asp-containing peptide overproduce the fibronectin receptor. J Cell Biol 105:1175–1182

Del Buono R, Pignatelli M, Bodmer WF, Wright NA (1991) The role of the arginine-glycine-aspartic acid-directed cellular binding to type I collagen and rat mesenchymal cells in colorectal tumour differentiation. Differentiation 46:97–103

D'Souza B, Berdichevsky F, Kyprianou N, Taylor-Papadimitrious J (1993) Collagen-induced morphogenesis and expression of the α_2-integrin subunit is inhibited in c-erbB2-transfected human mammary epithelial cells. Oncogene 8:1797–1806

Elices MJ, Hemler ME (1989) The human integrin VLA-2 is a collagen receptor on some cells and a collagen/laminin receptor on others. Proc Natl Acad Sci USA 86:9906–9910

Emerman JT, Pitelka DR (1977) Maintenance and induction of morphological differentiation in dissociated mammary epithelium on floating collagen membranes. In Vitro 13:316–323

Giancotti F, Ruoslahti E (1990) Elevated levels of the $\alpha_5\beta_1$ fibronectin receptor suppress the transformed phenotype of Chinese hamster ovary cells. Cell 60:349–350

Gui GPH, Wells CA, Browne PD, Yeomans P, Jordan S, Puddefoot JR, Vinson GP, Carpenter R (1995a) Integrin expression in primary breast cancer and its relation to axillary nodal status. Surgery 117: 102–108

Gui GPH, Puddefoot JR, Vinson GP, Wells CA, Carpenter R (1995b) Modulation of very late activation-2 laminin receptor function in breast cancer metastasis. Surgery 118:245–250

Hall PA, Coates P, Lemoine NR, Horton MA (1991) Characterization of integrin chains in normal and neoplastic human pancreas. J Pathol 165:33–41

Hemler ME, Crouse C, Sonnenberg A (1989) Association of the VLA α_6 subunit with a novel protein. J Biol Chem 264:6529–6535

Humphries MJ, Olden K, Yamada KM (1986) A synthetic peptide from fibronectin inhibits experimental metastasis of murine melanoma cells. Science 233:467–469

Hynes RO (1992) Integrins: versatility, modulation, and signaling in cell adhesion. Cell 69:11–25

Ignatius MJ, Reichardt LF (1988) Identification of a neuronal laminin receptor: an Mr 200k/120k integrin heterodimer that binds laminin in a divalent cation-dependent manner. Neuron 1:713–725

Ignatius MJ, Large TH, Houde M, Tawil JW, Barton A, Esch F, Carbonetto S, Reichardt LF (1990) Molecular cloning of the rat integrin α_1 subunit: a receptor for laminin and collagen. J Cell Biol 111:709–720

Kajiiji S, Tamura RN, Quaranta V (1989) A novel integrin (aEb4) from human epithelial cells suggests a fourth family of integrin adhesion receptors. EMBO 8:673–680

Keely PJ, Fong AM, Zutter MM, Santoro SA (1995) Alteration of collagen-dependent adhesion, motility, and morphogenesis by the expression of antisense α_2 integrin mRNA in mammary cells. J Cell Sci 108:595–607

Kennel SJ, Foote LJ, Flynn KM (1986) Tumor antigen on benign adenomas and on murine lung adenomas quantitated by a two-site monoclonal antibody assay. Cancer Res 46:707–712

Kern A, Eble J, Golbik R, Kuhn K (1993) Interaction of type IV collagen with the isolated integrins $\alpha_1\beta_1$ and $\alpha_2\beta_1$. Eur J Biochem 215:151–159

Kirchofer D, Languino LR, Ruoslahti E, Pierschbacher MD (1990) $\alpha_2\beta_1$ integrins from different cell types show different binding specificities. J Biol Chem 265:615–618

Klein CE, Dressel D, Steinmayer T, Mauch C, Eckes B, Krieg T, Bankert RB, Lueber L (1991a) Integrin $\alpha_2\beta_1$ is upregulated in fibroblasts and highly aggressive melanoma cells in three-dimensional collagen lattices and mediates the reorganization of collagen I fibrils. J Cell Biol 115:1427–1436

Klein CE, Steinmayer T, Kaufman D, Weber L, Broker E-B (1991b) Identification of a melanoma progression antigen as integrin VLA-2. J Invest Dermatol 96:281–284

Koretz K, Schlag P, Boumsell L, Moller P (1991) Expression of VLA-α_2, VLA-α_6, and VLA-β_1 chains in normal mucosa and adenomas of the colon, and in colon carcinomas and their liver metastases. Am J Pathol 138:741–750

Korhonen M, Laitinen L, Ylanne J, Koukolis GK, Quaranta V, Juusela H, Gould VE, Virtanen I (1992) Integrin distributions in renal cell carcinomas of various grades of malignancy. Am J Pathol 141:1161–1171

Koukoulis GK, Virtanen I, Korhonen M, Laitnen L, Quaranta V, Gould VE (1991) Immunohistochemical localization of integrins in the normal, hyperplastic and neoplastic breast: correlation with their functions as receptors and cell adhesion molecules. Am J Pathol 139:787–799

Koukoulis GK, Virtanen I, Moll R, Quaranta V, Gould VE (1993) Immunolocalization of integrins in the normal and neoplastic colonic epithelium. Virchows Arch B Cell Pathol 63:373–383

Kramer RH, Marks N (1989) Identification of integrin collagen receptors on human melanoma cells. J Biol Chem 264:4684–4688

Languino LR, Gehlsen KR, Wayner EA, Carter WG, Engvall E, Ruoslahti E (1989) Endothelial cells use $\alpha_2\beta_1$ integrin as a laminin receptor. J Cell Biol 109:2455–2462

Lauffenburger DA, Horwitz AF (1996) Cell migration: a physically integrated molecular process. Cell 84:359–369

Lee EYP, Lee W, Kaetzel CS, Parry G, Bissell MJ (1985) Interaction of mouse mammary epithelial cells with collagen substrata: regulation of casein gene expression and secretion. Proc Natl Acad Sci USA 82:1419–1423

Liebert M, Wedemeyer G, Stein JA, Washington RW Jr, Van Waes C, Carey TE (1993) The monoclonal antibody BQ16 identifies the alpha 6 beta 4 on bladder cancer. Hybridoma 12:67–80

Lindmark G, Gerdin B, Pahlman L, Glimelius B, Gehlsen K, Rubin K (1993) Interconnection of integrins alpha 2 and alpha 3 and structure of the basal membrane in colorectal cancer: relation to survival. Eur J Surg Oncol 19:50–60

Liu D, Gagliardi G, Nasim MM, Alison MR, Oates T, Lalani EA, Stamp GW, Pignatelli M (1994) TGF-a can act as morphogen and/or mitogen in a colon-cancer cell line. Int J Cancer 56:603–608

Mortarini R, Anchini A, Parmiani G (1991) Heterogeneity for integrin expression and cytokine-mediated VLA modulation can influence the adhesion of human melanoma cells to extracellular matrix proteins. Int J Cancer 47:551–559

Oka T, Yoshimura M, Lavanadero S, Wada K, Ohba Y (1991) Control of growth and differentiation of the mammary gland by growth factors. J Dairy Sci 74:2788–2800

Pignatelli M, Bodmer WF (1988) Genetics and biochemistry of collagen binding triggered glandular differentiation in a human colon carcinoma cell inc. Proc Natl Acad Sci USA 85:5561–5565

Pignatelli M (1990) Integrin cell adhesion molecules and colorectal cancer (editorial). J Pathol 162:95–97

Pignatelli M, Hanaby AM, Stamp GWH (1991a) Low expression of β_1, α_2, and α_3 subunits of VLA integrins in malignant mammary tumors. J Pathol 165:25–32

Pignatelli M, Smith MEF, Bodmer WF (1991b) Low expression of collagen receptors in moderate and poorly differentiated colorectal adenocarcinomas. Br J Cancer 61:636–638

Pignatelli M, Cardillo MR, Hanaby A, Stamp GW (1992) Integrins and their accessory adhesion molecules in mammary carcinomas: loss of polarization in poorly differentiated tumors. Human Pathol 23:1159–1166

Plantefaber LC, Hynes RO (1989) Changes in integrin receptors on oncogenetically transformed cells. Cell 56:281–290

Reichardt LF, Tomaselli KJ (1991) Extracellular matrix molecules and their receptors: function in neural development. Annu Rev Neurosci 14:531–570

Reichmann E, Ball R, Groner B, Friis RR (1989) New mammary epithelial and fibroblastic cell clones in coculture form structures competent to differentiate functionally. J Cell Biol 108:1127–1138

Ruoslahti E (1991) Integrins. J Clin Invest 87:1–5

Saelman EUM, Keely PJ, Santoro SA (1995) Loss of MDCK cell $\alpha_2\beta_1$ integrin expression results in reduced cyst formation, failure to hepatocyte growth factor/scatter factor-induced branching morphogenesis, and increased apoptosis. J Cell Sci 108:3531–3540

Santoro SA, Zutter MM (1995) The $\alpha_2\beta_1$ integrin: a collagen receptor on platelets and other cells. Thromb Haemost 74:813–821

Schiro JA, Chan BMC, Roswit WT, Kassner PD, Pentland AP, Hemler ME, Eisen AZ, Kupper TS (1991) Integrin $\alpha_2\beta_1$ (VLA-2) mediates reorganization and contraction of collagen matrices by human cells. Cell 87:403–410

Schreiner C, Fisher M, Hussein S, Juliano RL (1991) Increased tumorigenicity of fibronectin tumor-deficient Chinese hamster ovary cell variants. Cancer Res 51:1738–1740

Schwartz MA, Denninghoff K (1994) α_v integrins mediate the rise in intracellular calcium in endothelial cells on fibronectin even though they play a minor role in adhesion. J Biol Chem 269:11133–11137

Sonnenberg A, Linders CJT, Daams JH, Kennel SJ (1990) The $\alpha_6\beta_1$ (VLA-6) and $\alpha_6\beta_4$ protein complexes: tissue distribution and biochemical properties. J Cell Sci 96:207–217

Stallmach A, Von Lampe B, Matthes H, Bornhoft G, Riecken EO (1992) Diminished expression of integrin adhesion molecules on human colonic epithelial cells during the benign to malignant tumor transformation. Gut 33:342–346

Stamp GWH, Pignatelli M (1991) Distribution of β_1, α_2, and α_3 integrin chains in basal cell carcinomas. J Pathol 163:307–3113

Strange R, Li F, Friis RR, Reichmann E, Haenni B, Burri PH (1991) Mammary epithelial differentiation in vitro: minimum requirements for a functional response to hormonal stimulation. Cell Growth Differ 2:549–559

Tawil NJ, Houde M, Blacher R, Esch F, Reichardt LF, Turner DC, Carbonetto S (1990) $\alpha_1\beta_1$ integrin heterodimer functions as a dual laminin/collagen receptor in neural cells. Biochemistry 29:6540–6544

Taylor-Papadimitriou J, Berdichevsky F, D'Souza B, Burchill J (1993) Human models of breast cancer. Cancer Surv 16:59–78

Van Duinen CM, Van Den Broek LJ, Vermeer BJ, Fleuren GJ, Bruijn JA (1994) The distribution of cellular adhesion molecules in pigmented skin lesions. Cancer 73:2131–2139

Varner JA, Cheresh DA (1996) Integrins and cancer. Curr Opin Cell Biol 8:724–730

Weinel RJ, Rosendahl A, Neumann K, Chaloupka B, Erb O, Rothmund M, Santoro S (1992) Expression and function of VLA α_2, α_3, α_5, α_6 integrin receptors in pancreatic carcinoma. Int J Cancer 52:827–833

Werb Z, Tremble PM, Behrendsten O, Crowley E, Damsky CH (1989) Signal transduction through the fibronectin receptor induces collagenase and stromelysin gene expression. J Cell Biol 109:877–899

Ye J, Xu RH, Taylor-Papadimitriou J, Pitha PM (1996) Sp1 binding plays a critical role in the Erb-B2- and v-ras-mediated downregulation of alpha2-integrin expression in human mammary epithelial cells. Mol Cell Biol 16:6178–6189

Zutter MM, Santoro SA (1990) Widespread histologic distribution of the $\alpha_2\beta_1$ integrin cell surface collagen receptor. Am J Pathol 137:113–120

Zutter MM, Mazoujian G, Santoro SA (1990) Decreased expression of integrin adhesive protein receptors in adenocarcinoma of the breast. Am J Pathol 137:863–870

Zutter MM, Krigman HR, Santoro SA (1993) Altered integrin expression in adenocarcinoma of the breast. Analysis by in situ hybridization. Am J Pathol 142:1439–1448

Zutter MM, Santoro SA, Painter AS, Tsung YL, Gafford A (1994) The human α_2 integrin gene promoter: identification of positive and negative regulatory elements important for cell-type and developmentally-restricted gene expression. J Biol Chem 269:463–469

Zutter MM, Santoro SA, Staatz WD, Tsung YL (1995a) Re-expression of the $\alpha_2\beta_1$ integrin abrogates the malignant phenotype of breast carcinoma cells. Proc Natl Acad Sci USA 92:7411–7415

Zutter MM, Painter AS, Staatz WD, Tsung YL (1995b) Regulation of the α_2 integrin gene expression during megakaryocytic differentiation: a common theme of 3 necessary elements for megakaryocyte-specific, developmentally regulation gene expression. Blood 86:3006–3014

Zutter MM, Ryan EE, Painter AS (1997) Binding of phosphorylated Sp1 protein to tandem Sp1 binding sites regulates α_2 integrin gene core promoter activity. Blood 90:678–689

Subject Index

A
$\alpha 4\beta 1$ 67
$\alpha 4\beta 7$ 67
$\alpha E\beta 7$ 70
adhesion 33, 78, 172
- cascade 28
- molecules (AM) 66, 143, 147, 154, 161
- receptor 143
adoptive transfer of diabetes 72
afferent lymphatics 149–151
aggregate formation 156, 157
allograft 87
- rejection 85
alternative splicing 144, 162
antigen-presenting cells (APCs) 87
antigens, soluble 92
apoptosis 130, 151
arteriopathy, coronary 93
autoimmune disease, role of virus-induced bystander activation 118

B
B cell
- development 26
- immunity 93
blood-brain barrier (BBB) 51, 52
breast cancer 169
Burkitt's tumors 127

C
cancer (carcinoma) 136
- breast 169
- colorectal 171
carbohydrate moiety 159
cardiac myocytes 86
CD18-deficient mice 113
CD28 86
CD44 71, 144
- adhesive receptor 152
- isoforms 162
CD54 74
CD102 74
cell
- dissemination 156
- homing 143, 145, 152
- migration 143, 152
- motility 143
- proliferation 145, 151
central nervous system (CNS) 51–61, 76
chemokines 129
Chinese hamster ovary (CHO) cells 132, 135, 171
chondroitin sulfate 144
chronic inflammation 69
collagen 144, 157
colorectal cancer 171
corneas 90
coronary arteriopathy 93
costimulatory signals 86
CS-1 2–4
cytokines 75, 93
cytoplasmic domains 35

D
α4-deficient chimeric mice 26, 28
αE deficient mice 28
β1-deficient mice 26
β7-deficient mice 28
deglycosylation 158, 159
delayed-type hypersensitivity (DTH) 112
diagnosis 162
dissemination 154, 156, 160
DTH (delayed-type hypersensitivity) 112
Dukes stage 171

E
encephalitis 54
endothelial cells 66
enhancer 180
erb-B2 180
experimental autoimmune encephalomyelitis (EAE) 51–61, 67
experimental metastases 154
extracellular matrix (ECM) 2, 5, 143, 145, 158, 159, 167

F
fibronectin 68, 128, 144, 157
FK506 93
focal adhesion targeting (FAT) 9, 11
frozen section adhesion assay 57

G
gastrointestinal tract 27
α4 gene inactivation 25
gene target mice 26
gene targeting 24
N-glycosylation 158
glycoprotein 144, 158
glycosaminoglycan 158
graft arteriopathy 93, 95
gut associated lymphoid tissue 27
gylocoproteins 159

H
heterotypic aggregates (HA) 155, 156, 158–161
high endothelial venule (HEV) 28, 69, 143, 149, 150
- mucosal 147
homing 150–152, 161
homotypic adhesion 133, 151
hyaluranic acid 144, 157
hyaluronan 144, 145, 159, 161
hyaluronidase 158–161

I
ICAM-1 55, 74
ICAM-2 74
ICAM-deficient mice 113
IDDM (insulin-dependent diabetes mellitus) 67
IFN (*see* interferon)
IL (*see* interleukin)
immune
- response 33
- suppression 89
inflammation, chronic 69
inflammatory
- mediators 68
- response 109
inside-out signaling 33–46
insulin-dependent diabetes mellitus (IDDM) 67
insulitis 67
integrin adhesion receptors, β1 subfamily of 1
integrins 34, 143, 145, 154, 161, 162
- α4-integrins 23–30, 51–61, 125–138, 151
- β7-integrins 23–30
interferon (IFN)
- IFN-γ 75, 93
- role in bystander activation 104, 116
interleukin
- IL-1 68
- IL-2 93
- IL-4 68, 93
- IL-10 75, 93
intestinal intraepithelial lymphocytes 27
invasin 128
islets 67
isograft 87

K
β1 knockouts 26
β7 knockouts 26

L
lacrimal gland 76
lamina propria lymphocytes 27
laminin 144
LCMV (lymphocytic chriomeningitis) 100
LFA-1 (*see* lymphocyte function associated antigen-1)
lipopolysaccharide (LPS) 68
lymph nodes 66
- mesenteric 151
lymphatics, afferent 150, 151
lymphocyte function associated antigen-1 (LFA-1) 66, 86, 144, 145, 154, 156
- integrin 152, 155
lymphocyte 66
- migration 66
- traffic 90
lymphocytic chriomeningitis (LCMV) 100
lymphoid organs 152, 161
lymphoma 151, 154, 156, 158, 162
- cells 147, 159
- dissemination 154, 161
lymphotoxin 75

M
MAdCAM-1 (*see* mucosal addressin cell adhesion molecule-1)
major histocompatibility complex (MHC) 148, 150
- antigen 85
malignant lymphomatous polyposis (MLP) 132
mammary gland 174
matrix 33
MCP-1 (monocyte chemoattractant protein-1) 112
melanoma 126, 130, 133, 171
memory T lymphocytes 150
mesenteric lymph nodes 147, 151
metalloproteinase 133
metastasis 151, 158, 160, 161
- experimental 154
- spontaneous 154

metastatic
- cascade 160
- cells 147
- phenotype 150
- spread 154
MHC (*see* major histocompatibility complex)
migration 154, 160, 161
mitogenic signal 157
MLP (malignant lymphomatous polyposis) 132
morphogenesis 174
motility 176
mucosal addressin cell adhesion molecule-1 (MAdCAM-1) 28, 66, 128
multiple sclerosis 56
multistep paradigm 54, 61

N
neuraminidase 158, 159
non-*Hodgkin's* lymphomas 127, 133
nonobese diabetic (NOD) mouse 67

O
osteopontin 144, 157

P
p130 Cas 13–16
pancreas 66
paxillin 4, 6, 8, 10–12, 19
peripheral lymph node invasion 152
peripheral node addressin (PNAd) 66
Peyer's patches 27, 66, 145, 147, 151
phorbol ester 158–160
platelet-derived growth factor 95
pp105 12, 13, 15–17
pp105/Cas-L 19
pp125FAK 2, 4, 6–9, 11, 19
- paxillin-binding domain 9–11
primary growth 151
programmed cell death 130, 151
promoter 180
protein kinase C (PKC) 158

R
R-ras 129
RAG-1 deficient mice 25
RAG-2 deficient mice 25
ras 168
regulatory molecules 40
retrovirus-mediated gene transfer 148, 151
rhabdomyosarcoma 179
Rous sarcoma virus 168

S
salivary gland 72
sarcoma 127, 132
L-selectin 66
sialadenitis 76
sialic acid 158, 159
signal transduction 38
silencer 180
skin 87
- allografts 92
soluble antigens 92
spleen invasion 152
spontaneous
- aggregate formation 155
- metastases 154
src 168
Stamper Woodruff assay 28
synovial fibroblasts 133

T
T cell
- activation 101
- development 26
- lymphoma 135, 147, 148, 161
- migration 51–61, 109
- receptor 129
- response 101
Th1 93
Th2 93
therapy 162
thrombospondin 128
TM4SF (transmembrane-4 superfamily) 136
TNF-α (tumor necrosis factor-α) 68, 75
tolerance 86
transendothelial migration 143
tumor
- aggregation 156
- dissemination 156
- metastasis 125–138
- progression 154
tunicamycin 158
tyrosine phosphorylated proteins 5
tyrosine phosphorylation 3, 4, 6–8, 12, 14, 19

V
vascular cell adhesion molecule-1 (VCAM-1) 2, 3, 29, 55, 66, 128, 132
very late antigens (VLA) 1
- β1-integrins 2, 5, 6, 12, 14, 16, 19
- molecules 2
VLA-4 2–5, 145, 144, 151
vessels 76
vinculin 9–11
virus-specific CTLs, phenotype of 106

Printing: Saladruck, Berlin
Binding: Buchbinderei Lüderitz & Bauer, Berlin

Current Topics in Microbiology and Immunology

Volumes published since 1989 (and still available)

Vol. 190: **Koprowski, Hilary; Lipkin, W. Ian (Eds.):** Borna Disease. 1995. 33 figs. IX, 134 pp. ISBN 3-540-57388-7

Vol. 191: **ter Meulen, Volker; Billeter, Martin A. (Eds.):** Measles Virus. 1995. 23 figs. IX, 196 pp. ISBN 3-540-57389-5

Vol. 192: **Dangl, Jeffrey L. (Ed.):** Bacterial Pathogenesis of Plants and Animals. 1994. 41 figs. IX, 343 pp. ISBN 3-540-57391-7

Vol. 193: **Chen, Irvin S. Y.; Koprowski, Hilary; Srinivasan, Alagarsamy; Vogt, Peter K. (Eds.):** Transacting Functions of Human Retroviruses. 1995. 49 figs. IX, 240 pp. ISBN 3-540-57901-X

Vol. 194: **Potter, Michael; Melchers, Fritz (Eds.):** Mechanisms in B-cell Neoplasia. 1995. 152 figs. XXV, 458 pp. ISBN 3-540-58447-1

Vol. 195: **Montecucco, Cesare (Ed.):** Clostridial Neurotoxins. 1995. 28 figs. XI., 278 pp. ISBN 3-540-58452-8

Vol. 196: **Koprowski, Hilary; Maeda, Hiroshi (Eds.):** The Role of Nitric Oxide in Physiology and Pathophysiology. 1995. 21 figs. IX, 90 pp. ISBN 3-540-58214-2

Vol. 197: **Meyer, Peter (Ed.):** Gene Silencing in Higher Plants and Related Phenomena in Other Eukaryotes. 1995. 17 figs. IX, 232 pp. ISBN 3-540-58236-3

Vol. 198: **Griffiths, Gillian M.; Tschopp, Jürg (Eds.):** Pathways for Cytolysis. 1995. 45 figs. IX, 224 pp. ISBN 3-540-58725-X

Vol. 199/I: **Doerfler, Walter; Böhm, Petra (Eds.):** The Molecular Repertoire of Adenoviruses I. 1995. 51 figs. XIII, 280 pp. ISBN 3-540-58828-0

Vol. 199/II: **Doerfler, Walter; Böhm, Petra (Eds.):** The Molecular Repertoire of Adenoviruses II. 1995. 36 figs. XIII, 278 pp. ISBN 3-540-58829-9

Vol. 199/III: **Doerfler, Walter; Böhm, Petra (Eds.):** The Molecular Repertoire of Adenoviruses III. 1995. 51 figs. XIII, 310 pp. ISBN 3-540-58987-2

Vol. 200: **Kroemer, Guido; Martinez-A., Carlos (Eds.):** Apoptosis in Immunology. 1995. 14 figs. XI, 242 pp. ISBN 3-540-58756-X

Vol. 201: **Kosco-Vilbois, Marie H. (Ed.):** An Antigen Depository of the Immune System: Follicular Dendritic Cells. 1995. 39 figs. IX, 209 pp. ISBN 3-540-59013-7

Vol. 202: **Oldstone, Michael B. A.; Vitković, Ljubiša (Eds.):** HIV and Dementia. 1995. 40 figs. XIII, 279 pp. ISBN 3-540-59117-6

Vol. 203: **Sarnow, Peter (Ed.):** Cap-Independent Translation. 1995. 31 figs. XI, 183 pp. ISBN 3-540-59121-4

Vol. 204: **Saedler, Heinz; Gierl, Alfons (Eds.):** Transposable Elements. 1995. 42 figs. IX, 234 pp. ISBN 3-540-59342-X

Vol. 205: **Littman, Dan R. (Ed.):** The CD4 Molecule. 1995. 29 figs. XIII, 182 pp. ISBN 3-540-59344-6

Vol. 206: **Chisari, Francis V.; Oldstone, Michael B. A. (Eds.):** Transgenic Models of Human Viral and Immunological Disease. 1995. 53 figs. XI, 345 pp. ISBN 3-540-59341-1

Vol. 207: **Prusiner, Stanley B. (Ed.):** Prions Prions Prions. 1995. 42 figs. VII, 163 pp. ISBN 3-540-59343-8

Vol. 208: **Farnham, Peggy J. (Ed.):** Transcriptional Control of Cell Growth. 1995. 17 figs. IX, 141 pp. ISBN 3-540-60113-9

Vol. 209: **Miller, Virginia L. (Ed.):** Bacterial Invasiveness. 1996. 16 figs. IX, 115 pp. ISBN 3-540-60065-5

Vol. 210: **Potter, Michael; Rose, Noel R. (Eds.):** Immunology of Silicones. 1996. 136 figs. XX, 430 pp. ISBN 3-540-60272-0

Vol. 211: **Wolff, Linda; Perkins, Archibald S. (Eds.):** Molecular Aspects of Myeloid Stem Cell Development. 1996. 98 figs. XIV, 298 pp. ISBN 3-540-60414-6

Vol. 212: **Vainio, Olli; Imhof, Beat A. (Eds.):** Immunology and Developmental Biology of the Chicken. 1996. 43 figs. IX, 281 pp. ISBN 3-540-60585-1

Vol. 213/I: **Günthert, Ursula; Birchmeier, Walter (Eds.):** Attempts to Understand Metastasis Formation I. 1996. 35 figs. XV, 293 pp. ISBN 3-540-60680-7

Vol. 213/II: **Günthert, Ursula; Birchmeier, Walter (Eds.):** Attempts to Understand Metastasis Formation II. 1996. 33 figs. XV, 288 pp. ISBN 3-540-60681-5

Vol. 213/III: **Günthert, Ursula; Schlag, Peter M.; Birchmeier, Walter (Eds.):** Attempts to Understand Metastasis Formation III. 1996. 14 figs. XV, 262 pp. ISBN 3-540-60682-3

Vol. 214: **Kräusslich, Hans-Georg (Ed.):** Morphogenesis and Maturation of Retroviruses. 1996. 34 figs. XI, 344 pp. ISBN 3-540-60928-8

Vol. 215: **Shinnick, Thomas M. (Ed.):** Tuberculosis. 1996. 46 figs. XI, 307 pp. ISBN 3-540-60985-7

Vol. 216: **Rietschel, Ernst Th.; Wagner, Hermann (Eds.):** Pathology of Septic Shock. 1996. 34 figs. X, 321 pp. ISBN 3-540-61026-X

Vol. 217: **Jessberger, Rolf; Lieber, Michael R. (Eds.):** Molecular Analysis of DNA Rearrangements in the Immune System. 1996. 43 figs. IX, 224 pp. ISBN 3-540-61037-5

Vol. 218: **Berns, Kenneth I.; Giraud, Catherine (Eds.):** Adeno-Associated Virus (AAV) Vectors in Gene Therapy. 1996. 38 figs. IX,173 pp. ISBN 3-540-61076-6

Vol. 219: **Gross, Uwe (Ed.):** Toxoplasma gondii. 1996. 31 figs. XI, 274 pp. ISBN 3-540-61300-5

Vol. 220: **Rauscher, Frank J. III; Vogt, Peter K. (Eds.):** Chromosomal Translocations and Oncogenic Transcription Factors. 1997. 28 figs. XI, 166 pp. ISBN 3-540-61402-8

Vol. 221: **Kastan, Michael B. (Ed.):** Genetic Instability and Tumorigenesis. 1997. 12 figs.VII, 180 pp. ISBN 3-540-61518-0

Vol. 222: **Olding, Lars B. (Ed.):** Reproductive Immunology. 1997. 17 figs. XII, 219 pp. ISBN 3-540-61888-0

Vol. 223: **Tracy, S.; Chapman, N. M.; Mahy, B. W. J. (Eds.):** The Coxsackie B Viruses. 1997. 37 figs. VIII, 336 pp. ISBN 3-540-62390-6

Vol. 224: **Potter, Michael; Melchers, Fritz (Eds.):** C-Myc in B-Cell Neoplasia. 1997. 94 figs. XII, 291 pp. ISBN 3-540-62892-4

Vol. 225: **Vogt, Peter K.; Mahan, Michael J. (Eds.):** Bacterial Infection: Close Encounters at the Host Pathogen Interface. 1998. 15 figs. IX, 169 pp. ISBN 3-540-63260-3

Vol. 226: **Koprowski, Hilary; Weiner, David B. (Eds.):** DNA Vaccination/Genetic Vaccination. 1998. 31 figs. XVIII, 198 pp. ISBN 3-540-63392-8

Vol. 227: **Vogt, Peter K.; Reed, Steven I. (Eds.):** Cyclin Dependent Kinase (CDK) Inhibitors. 1998. 15 figs. XII, 169 pp. ISBN 3-540-63429-0

Vol. 228: **Pawson, Anthony I. (Ed.):** Protein Modules in Signal Transduction. 1998. 42 figs. IX, 368 pp. ISBN 3-540-63396-0

Vol. 229: **Kelsoe, Garnett; Flajnik, Martin (Eds.):** Somatic Diversification of Immune Responses. 1998. 38 figs. IX, 221 pp. ISBN 3-540-63608-0